KU-184-551

Contents

Preface

These notes report on recent progress in the determination of automorphism groups of compact Riemann surfaces. Already in the nineteenth century, automorphisms of Riemann surfaces were studied. In this context, an automorphism of a Riemann surface means a biholomorphic self-equivalence, often this is also called a conformal automorphism. In 1893, A. Hurwitz proved in [**Hur93**] that the full automorphism group of a compact Riemann surface of genus $g \geq 2$ is finite, its order being bounded by $84(g-1)$. Two years later, the maximal order of a single automorphism was considered by A. Wiman in [**Wim95**]. More than half a century later, this question was revisited in [**Har66**], this time with emphasis on determining the minimal genus larger than or equal to 2 of a compact Riemann surface admitting an automorphism of prescribed order. The same question was treated for abelian groups in [**Mac65**] and for simple groups, in particular sporadic simple groups, in several publications, e.g. [**Wol89b, Wol89a, CWW92**].

If a group is concretely given as a group of symmetries of a mathematical object and not only abstractly via its isomorphism type, usually results about this group can be refined using representation theoretic methods. A natural representation of the automorphism group of a compact Riemann surface X is that on the complex vector space $\mathcal{H}^1(X)$ of holomorphic abelian differentials on X (see [**FK92**, Chapter V.2]). From this viewpoint, it is quite natural for us to focus our interest on the following two problems.

PROBLEM 1: Classify all groups of automorphisms of compact Riemann surfaces X of fixed genus $g(X) \geq 2$, up to equivalence of the natural action on $\mathcal{H}^1(X)$.

PROBLEM 2: Classify all characters of a given finite group G that arise from the natural action of groups of automorphisms (isomorphic to G) of compact Riemann surfaces X on $\mathcal{H}^1(X)$.

Both problems were treated in several publications by A. Kuribayashi, I. Kuribayashi, and H. Kimura, see [**Kur66, Kur83, Kur84, Kur86, Kur87, KK90a, KK90b, KK91, Kim93**]. We follow their approach and develop it further.

That is, for Problem 1, by the above mentioned bound on the group order, we first note that only finitely many groups must be considered for fixed genus $g \geq 2$. Then we use the fact that a group of automorphisms of a compact Riemann surface of fixed genus occurs as epimorphic image of one of a finite number of finitely presented groups. In the appendix of [**KK90b**], this is expressed by the statement that 'in each genus, there is a finite procedure for deciding which groups arise with $GL_g(\mathbb{C})$-conjugacy from Riemann surfaces'. However, instead of adhoc group theoretic arguments as in [**KK90a, KK90b**], we use a recently obtained classification of groups of small order (see [**BE99**]) or alternatively a recursive construction algorithm to solve Problem 1 for small genera. It should be noted that the classification up to equivalence of the action on $\mathcal{H}^1(X)$ is a refinement of the classification up to isomorphism, and that it means a classification of the characters corresponding to the natural actions on the spaces $\mathcal{H}^1(X)$.

In this sense, Problem 2 is dual to Problem 1. The idea for attacking this problem is to consider certain necessary conditions on a character to arise from an action as required. These conditions are derived from the description of such a character by means of the fixed points of the action on X, a description given by the Eichler Trace Formula. In [**KK91**], it is observed that only very few characters of the general linear group $GL_3(2)$ satisfy these necessary conditions but do *not* arise from natural actions on some $\mathcal{H}^1(X)$. The authors note that 'this phenomenon seems to be rooted in some structure of groups although we cannot explicitly point out which'. We confirm the observation in a more systematic treatment of several classes of groups, including a study of the possibility of infinitely many exceptional characters, as occurs for example in the case of the nonabelian groups of order 8 (see [**Kim93**]).

Each chapter will have its own introductory paragraph. Here we just summarize the contents of the chapters.

In the first three chapters, we introduce briefly the theory that is needed later on. More specifically, **Chapter 1** collects some facts about Riemann surfaces; it is not our aim to give a self-contained introduction to this theory, rather we refer to standard textbooks, e.g. [**Leh64, FK92**], for proofs of classical results. **Chapter 2** first lists the – in fact quite few – needed results in the character theory of finite groups –proofs can be looked up for example in [**Isa76**]. We then deal with the special case we are interested in, the character of the natural action of $G \leq \text{Aut}(X)$ on $\mathcal{H}^1(X)$, for a compact Riemann surface X. **Chapter 3** describes, after studying the example of abelian groups of automorphisms, the relation of this G-character and the fixed points of G on X.

In our situation, a given G-character arises from an action on $\mathcal{H}^1(X)$ for some X if and only if a so-called surface kernel epimorphism onto G exists;

for given G, this is an epimorphism from one of a finite number of finitely presented groups Γ (for details, see Section 3.8). Asking whether such an epimorphism exists means to ask whether a homomorphism exists and if yes whether it is surjective. Both questions are treated in **Chapter 4**, character theoretic arguments such as the interpretation of structure constants and the inspection of the relations between the character tables of a group and its (maximal) subgroups play the main role here. The first section of Chapter 4 lists the necessary tools. They are well-known from problems such as to determine the Hurwitz groups of given order or the strong symmetric genus of a given group (see [**Con85, Con91, CWW92, Tuc83, Wol89b, Wol89a**]), and from related problems concerning the inverse problem of Galois theory (see [**Mat87**]). As an application to Riemann surfaces, the next section shows –for most of the sporadic simple groups as examples– how good bounds for the strong symmetric genus of a group can be computed with character theoretic methods; the library of character tables available in the computer algebra system GAP (see [**S+94**]) is an important source of information for that task. In the last section of Chapter 4, the obtained conditions on the existence of surface kernel epimorphisms are first reformulated in terms of so-called signatures, and then stronger conditions are derived for the case that the group in question is solvable.

With these prerequisites, we are able to present in **Chapter 5** a classification of all groups of automorphisms of compact Riemann surfaces of genus g, as required in Problem 1, for $2 \leq g \leq 48$. In earlier publications, only classifications for $2 \leq g \leq 5$ had been achieved (see [**Kur84, Kur86, KK90a, KK90b**]); an important reason why our methods reach farther than those used there (and in particular farther than those that were available at Hurwitz's time) is that we make heavy use of computer calculations. Chapter 5 describes an algorithm for enumerating the characters in question, and shows several examples. A byproduct is the classification of *irreducible* actions on $\mathcal{H}^1(X)$, the complete list is given in Appendix B.

Then we consider Problem 2, first for the easy case of groups with real character table in **Chapter 6**, and then –motivated by these results– in **Chapter 7** for the general case. The used necessary conditions on a character to be induced by an action of a group of automorphisms of a compact Riemann surface X on the vector space $\mathcal{H}^1(X)$ are formulated as linear equations and inequalities in terms of the decomposition of the given character into irreducibles. These conditions turn out to be sufficient if the multiplicity of the trivial character as a constituent is large enough. For the remaining characters, and hence for the solution of Problem 2 for the given group G, say, the answer depends on G. More specifically, there are groups for which infinite series of characters can be identified that satisfy the necessary conditions but are not induced by the action of a group of automorphisms of a compact

Riemann surface X on $\mathcal{H}^1(X)$; and there are groups for which one can prove that at most finitely many characters satisfying the necessary conditions fail to come from an action on a compact Riemann surface. For example, many finite simple groups are of the latter kind, and the check for this involves only computations with the character table of the given group; if the generic character table of a series of groups is available then this table suffices to perform this check for the finiteness of the set of exceptions for each group in the series. Finally, a full solution of Problem 2 for given (small) groups can be obtained with algorithms that are also presented in Chapters 6 and 7.

Although our arguments are completely elementary, the results are again beyond prior publications. Problem 2 had been considered for the nonabelian groups of order 8 (see [Kim93]), all cyclic groups (see [Kur87]), and the group $GL_3(2)$ (see [KK91]). We generalize this to full classifications for arbitrary dihedral groups (see Section 26), abelian groups (see Section 32), several specific examples such as small symmetric, alternating, and linear groups (see Sections 27.3, 29.3), and partial results for several infinite series of simple and nearly simple groups (see Sections 27.1, 28.1, 29.1, 35.1). Additionally, we simplify the proofs of the known results, mainly due to the concepts introduced in Sections 22 and 30.

A number theoretic problem involving Maillet's Determinant and its generalizations (see [CO55, Küh79, Tat82]) arises in Chapter 7. It is studied in Appendix C.

We do *not* consider possible further generalizations, e.g., to study spaces $\mathcal{H}^q(X)$ of differentials of order $q > 1$ (see [Kur87, FK92]), automorphism groups of bordered surfaces (see [BEGG90]), or higher-dimensional manifolds.

The computational results were obtained with the help of the computer algebra system GAP (see [S⁺94]). A GAP database of the automorphism groups of genera from 2 to 48 is freely available from the author as well as the programs that were used in the computations.

This work was initiated by Prof. Herbert Pahlings. I would like to thank him for his interest in the project. Furthermore I thank Klaus Lux for many discussions. Finally, I owe gratitude to Lehrstuhl D für Mathematik at RWTH Aachen, whose computer equipment I was allowed to use.

Aachen, July 1998   Thomas Breuer

Notation

The standard language and notation of set theory are used, disjoint union of sets A, B is denoted by $A \uplus B$. Besides that, we use $\mathbb{Z}$, $\mathbb{Q}$, $\mathbb{R}$, and $\mathbb{C}$ to denote the ring of integers, and the fields of rational, real, and complex numbers, respectively.

We write ζ_m for the primitive complex m-th root of unity $e^{2\pi i/m}$.

The composition of mappings is written either as $\varphi_2 \circ \varphi_1$, with $(\varphi_2 \circ \varphi_1)(x) = \varphi_2(\varphi_1(x))$, or as $\varphi_1 \varphi_2$, with $x(\varphi_1 \varphi_2) = (x\varphi_1)\varphi_2 = \varphi_2(\varphi_1(x))$.

G will denote a finite group, $G^\times$ the set of nonidentity elements in G, and G' the derived subgroup of G.

For an element $h \in G$ (note that the letter g will be reserved to denote the genus of compact surfaces), $\langle h \rangle$ is the subgroup of G generated by h, and $|h| = |\langle h \rangle|$ is the *order* of h. $[G{:}H] = |G|/|H|$ is the *index* of the subgroup $H \leq G$ in G, $C_G(H) = \{\sigma \in G \mid \sigma^{-1}h\sigma = h \text{ for all } h \in H\}$ the *centralizer* of H in G, $N_G(H) = \{\sigma \in G \mid \sigma^{-1}H\sigma = H\}$ the *normalizer* of H in G. We will write $C_G(h)$ instead of $C_G(\langle h \rangle)$ for an element $h \in G$. A subgroup H is called *normal* if $N_G(H) = G$.

For two groups G and H, $G{:}\,H$ denotes the split extension of G with H. For example, $\mathbb{Z}^2{:}\,m$ is a group with a free abelian normal subgroup $\mathbb{Z}^2$ of rank 2 that has a cyclic group of order m as a complement.

Two elements or subgroups H, K of G are *conjugate* if there is $\sigma \in G$ with $\sigma^{-1}H\sigma = K$, we write $H \sim_G K$ in this situation. For a set S on which G acts via conjugation, $S/\sim_G$ denotes a set of representatives of the conjugacy classes.

The image of the point x under the group element σ is written as $x\sigma$ if the operation is not conjugation, the *orbit* $\{x\sigma \mid \sigma \in G\}$ of the point x under this action of G is written as xG.

S_n and A_n denote the *symmetric* and *alternating* groups respectively on n symbols, D_{2n} the *dihedral group* of order $2n$, Q_8 the quaternion group of order 8. Finally, $GL_n(F)$, $SL_n(F)$, $PGL_n(F)$, and $PSL_n(F)$ denote the general linear group, the special linear group, the projective general linear group, and the projective special linear group, respectively, in dimension n over the field F (if F is a field) or over a field with F elements (if F is an integer). Instead of $PSL_n(F)$, we also write $L_n(F)$.

For more special notations such as $\mathrm{Fix}_X(h)$ or $I(m)$, the definition in the text can be found via the index at the end of the book.

Names of conjugacy classes and characters refer either to character tables shown in the text or to character tables in [CCN$^+$85]; for example, the class 2A of the simple group $L_2(7)$ is the class of involutions, and the classes 7A and 7B are the two classes of elements of order 7, as given in [CCN$^+$85, p. 3].

CHAPTER 1

Compact Riemann Surfaces

For our purposes, it is sufficient to think of Riemann surfaces as topological spaces with additional structure, and this is defined in Section 1.

In Section 2, the necessary terms and facts from algebraic topology are collected. We will need mainly the notions of smooth and branched coverings, universal coverings, and orbit spaces. See [**Mas91**, Chapter V] for a more general and more detailed introduction to these topics. The statements and proofs for the special case that the topological spaces are Riemann surfaces can also be found in [**For77**, Kap. I, §§ 3–5].

We are interested in compact Riemann surfaces of genus at least 2, they arise as orbit spaces $\mathcal{U}/K$ of the upper half plane $\mathcal{U}$ by Fuchsian groups K. These groups are studied in Section 3. The presentation of the material follows [**Leh64**] and [**JS87**].

Some useful statements about the Riemann–Hurwitz Formula can be found in Section 4, and Section 5 sketches two examples.

1. Basic Facts about Riemann Surfaces

Recall that a topological space X is *connected* if it is not the union of two disjoint, nonempty, open sets. X is *Hausdorff* if, for every two distinct points x_1, x_2 in X, there are disjoint neighborhoods of x_1 and x_2. A map $\varphi \colon X \to Y$ is called *continuous* if for each open set $U \subseteq Y$ the preimage $\varphi^{-1}(U)$ is open in X, and φ is called a *homeomorphism* (or *topological equivalence*) if it is continuous and bijective, and its inverse is also continuous.

A connected Hausdorff space X with the property that each point $x \in X$ has a neighborhood that is homeomorphic to an open subset of $\mathbb{R}^n$ (with its natural topology) is called an *n-manifold*. A 2-manifold is also called a *surface*. By identifying the topological spaces $\mathbb{R}^2$ and $\mathbb{C}$, each point of a surface X has a neighborhood U that is homeomorphic to an open subset V of $\mathbb{C}$.

Such a homeomorphism $\varphi \colon U \to V$ is called a *local coordinate*, and the pair (U, φ) is called a *complex chart* on X. For any two complex charts (U_1, φ_1),

1

(U_2, φ_2) with $U_1 \cap U_2 \neq \emptyset$, the coordinate change

$$(\varphi_1^{-1}\varphi_2)|_{\varphi_1(U_1 \cap U_2)} \colon \varphi_1(U_1 \cap U_2) \to \varphi_2(U_1 \cap U_2)$$

is a complex function. It is *holomorphic* (or *analytic* or *regular*) if it is complex differentiable at every point of $\varphi_1(U_1 \cap U_2)$. A holomorphic and bijective map is called *biholomorphic*, its inverse is automatically holomorphic. If $U_1 \cap U_2$ is empty or if the coordinate change is biholomorphic then the charts (U_1, φ_1) and (U_2, φ_2) are called *compatible*.

An *analytic atlas* on X is a system $\mathcal{A} = \{(U_i, \varphi_i) \mid i \in I\}$ of pairwise compatible charts with $\bigcup_{i \in I} U_i = X$. Two atlases $\mathcal{A}$, $\mathcal{A}'$ on X are *compatible* if each chart of $\mathcal{A}$ is compatible with each chart of $\mathcal{A}'$, or equivalently if $\mathcal{A} \cup \mathcal{A}'$ is also an atlas on X. Compatibility of analytic atlases is an equivalence relation, the equivalence classes are called *complex structures* on X.

A *Riemann surface* is a surface X together with a complex structure on X.

REMARK 1.1. Some authors do not require Riemann surfaces to be connected (see [**JS87**, pp. 167 ff.]), some authors require Riemann surfaces to be *second countable*, that is, there is a countable basis for its topology (see [**Mir95**, Def. I.1.18]).

Having defined the objects of our interest, we specify the appropriate class of structure preserving maps between Riemann surfaces X and Y as follows. A map $f \colon X \to Y$ is called *holomorphic* if f is continuous, and for each pair of charts $\varphi_1 \colon U_1 \to V_1$ on X and $\varphi_2 \colon U_2 \to V_2$ on Y with $f(U_1) \subseteq U_2$ the map $\varphi_1^{-1} f \varphi_2 \colon V_1 \to V_2$ is holomorphic as a complex function; f is called *biholomorphic* if f is bijective and holomorphic. As for complex functions, the inverse of a biholomorphic map is holomorphic.

If a biholomorphic map $f \colon X \to Y$ exists then X and Y are considered as indistinguishable, and they are called *conformally equivalent* (or *isomorphic as Riemann surfaces*). By definition, conformally equivalent Riemann surfaces are homeomorphic, so the equivalence relation defined by conformal equivalence is a refinement of topological equivalence.

A biholomorphic map $X \to X$ is called a *conformal automorphism* of X. The set of all conformal automorphisms of X is denoted by $\mathrm{Aut}(X)$, it is a group with respect to composition of maps, and is called the *full automorphism group* of X.

REMARK 1.2. Note that the term *automorphism* is used for other kinds of maps in other contexts. For a topological space, any homeomorphism is usually called a *topological automorphism*, and for a covering space (see Section 2), the group of covering transformations is also called the automorphism group of the covering. Besides that, there are of course the obvious notions of group automorphisms, field automorphisms etc., so $\mathrm{Aut}(G)$ denotes the automorphism group of the group G. These ambiguities should not cause confusion in the following.

EXAMPLE 1.3. Examples of Riemann surfaces are the complex plane $\mathbb{C}$, the Riemann sphere $\hat{\mathbb{C}} = \mathbb{C} \cup \{\infty\}$, the upper half plane $\mathcal{U} = \{z \in \mathbb{C} \mid \operatorname{Im}(z) > 0\}$, and the unit disk $\mathcal{D} = \{z \in \mathbb{C} \mid z \cdot \bar{z} \leq 1\}$. The analytic structure for $\mathbb{C}$, $\mathcal{U}$, and $\mathcal{D}$ is given by the identity map, the one for $\hat{\mathbb{C}}$ can be defined by the two charts

$$\varphi_1 \colon \begin{array}{ccc} \mathbb{C} & \to & \mathbb{C} \\ z & \mapsto & z \end{array} \quad \text{and} \quad \varphi_2 \colon \begin{array}{ccc} \hat{\mathbb{C}} \setminus \{0\} & \to & \mathbb{C} \\ z & \mapsto & 1/z. \end{array}$$

The Riemann surfaces $\mathcal{U}$ and $\mathcal{D}$ are conformally equivalent via the holomorphic maps

$$\varphi \colon \begin{array}{ccc} \mathcal{U} & \to & \mathcal{D} \\ z & \mapsto & (z-i)/(z+i) \end{array} \quad \text{and} \quad \varphi^{-1} \colon \begin{array}{ccc} \mathcal{D} & \to & \mathcal{U} \\ z & \mapsto & (-iz-i)/(z-1), \end{array}$$

see [**JS87**, p. 199]. We can visualize φ as the composition of the rotation $z \mapsto (z-i)/(-iz+1)$ of the Riemann sphere around the axis through ± 1, mapping 0 to $-i$, and the rotation $z \mapsto -iz$ of $\mathcal{D}$ around the origin.

$\mathbb{C}$ and $\mathcal{D}$ are homeomorphic via the map $\mathcal{D} \to \mathbb{C}$, $z \mapsto z/(1-|z|)$, but they are not conformally equivalent because each holomorphic map $\mathbb{C} \to \mathcal{D}$ is bounded and hence constant by Liouville's Theorem [**JS87**, Theorem A.4]. ⬦

Every transformation of the form $z \mapsto (az+b)/(cz+d)$ for complex numbers a, b, c, d with $ad-bc = 1$ is a conformal automorphism of $\hat{\mathbb{C}}$, and its identification with the pair of matrices

$$\pm \begin{pmatrix} a & b \\ c & d \end{pmatrix}$$

defines an isomorphism to the group $PSL_2(\mathbb{C})$. The sets $\mathbb{C}$, $\mathcal{U}$, and $\mathcal{D}$ are subsets of $\hat{\mathbb{C}}$, and the sets of those transformations that fix the point ∞, the circle $\mathbb{R} \cup \infty$, and the unit circle respectively are conformal automorphisms of these Riemann surfaces. Moreover, these are already the full groups of automorphisms.

THEOREM 1.4.

$$\operatorname{Aut}(\hat{\mathbb{C}}) = PSL_2(\mathbb{C}),$$
$$\operatorname{Aut}(\mathbb{C}) = \{z \mapsto az + b \mid a, b \in \mathbb{C}, a \neq 0\},$$
$$\operatorname{Aut}(\mathcal{U}) = PSL_2(\mathbb{R}).$$

Proof. [**JS87**, Theorem 4.17.3] □

Every compact Riemann surface X is homeomorphic to the g-holed torus, for a unique nonnegative integer g, see [**JS87**, Theorem 4.16.1], [**Mir95**,

Proposition I.1.23], [**For77**, 19.14]. The number g is called the *genus* of X, we will denote it by $g(X)$. For example, we have $g(\hat{\mathbb{C}}) = 0$, and $\hat{\mathbb{C}}$ is the unique compact Riemann surface of genus 0.

2. Basic Algebraic Topology

2.1. Branched Coverings. As above, let X and Y be topological spaces. A subset $A \subseteq X$ is called *discrete* if each point $a \in A$ has a neighborhood V in X such that $V \cap A = \{a\}$. A map $\varphi \colon X \to Y$ is called *discrete* if $\varphi^{-1}(y)$ is a discrete set in X for each $y \in Y$, and φ is called *open* if $\varphi(U)$ is open in Y for each open set $U \subseteq X$.

A continuous, open, and discrete map $\varphi \colon X \to Y$ is called a *branched covering*. In this case, the set $\varphi^{-1}(y)$ is called the *fibre* of $y \in Y$.

The point $x \in X$ is called a *ramification point* if there is no neighborhood U of x such that the restriction $\varphi|_U$ is injective. The image $\varphi(x)$ of a ramification point is called a *branch point*.

A branched covering map is called *unbranched covering* or *smooth covering* if it has no branch points. Smooth coverings can be characterized as local homeomorphisms, in the sense that a map $\varphi \colon X \to Y$ is a smooth covering if and only if each $x \in X$ has a neighborhood U such that $\varphi|_U \colon U \to \varphi(U)$ is a homeomorphism, with $\varphi(U)$ open in Y (see [**For77**, 4.4]).

2.2. Universal Coverings. Let I be the interval $[0, 1] \subseteq \mathbb{R}$. A *path* on X is a continuous map $I \to X$. The space X is called *path-connected* (or *arcwise-connected*) if for any two points x_0, $x_1 \in X$ a path u on X with $u(0) = x_0$ and $u(1) = x_1$ exists. A *loop* based at $x_0 \in X$ is a path on X such that $u(0) = u(1) = x_0$. Two loops u, $v \colon I \to X$ based at x_0 are called *homotopic* if there is a continuous map $F \colon I \times I \to X$ such that $F(t, 0) = u(t)$ and $F(t, 1) = v(t)$ for all $t \in I$, and $F(0, s) = F(1, s) = x_0$ for all $s \in I$.

X is called *simply connected* if X is path-connected and for any $x \in X$, any loop based at x is homotopic to the constant path $I \to X$, $t \mapsto x$. If X is simply connected then a smooth covering $\varphi \colon X \to Y$ has the following universal property (see [**Mas91**, Theorem V.5.1]).

For each connected, smooth covering $\psi \colon Z \to Y$ and each choice of $x_0 \in X$ and $z_0 \in Z$ with $\varphi(x_0) = \psi(z_0)$ there is a unique continuous map $f \colon X \to Z$ with $f(x_0) = z_0$ and $\varphi = f\psi$. Roughly speaking, φ factors through Z.

A smooth covering with this property is called a *universal covering*, if it exists then it is unique up to homeomorphism.

THEOREM 2.1. *Each Riemann surface X has a simply connected universal covering $\varphi \colon \tilde{X} \to X$, where $\tilde{X}$ is again a Riemann surface and φ is holomorphic.*

The universal covering of any Riemann surface is conformally equivalent to exactly one of $\hat{\mathbb{C}}$, $\mathbb{C}$, $\mathcal{U}$.

Proof. [**JS87**, Theorem 4.17.2]   □

2.3. Covering Transformations. Let $\varphi \colon X \to Y$ be a smooth covering. A *covering transformation* of φ is a homeomorphism $\sigma \colon X \to X$ with the property that $\sigma\varphi = \varphi$. With composition as multiplication, the covering transformations of φ form a group, which is denoted by $\mathrm{Aut}_Y(X)$. For path-connected X, this group acts fixed point freely on each fibre, see [**Mas91**, V.6.1]. The smooth covering φ is called *regular* (or a *Galois covering*) if $\mathrm{Aut}_Y(X)$ acts transitively on each fibre.

THEOREM 2.2. *A universal covering is regular.*

Proof. [**Mas91**, Lemma V.8.1]   □

2.4. Orbit Spaces. For a group G of homeomorphisms of a topological space X, the *canonical projection* onto the *orbit space* X/G is defined to map each point $x \in X$ to its orbit xG under the action of G.

If $X \to Y$ is a smooth covering and $G \leq \mathrm{Aut}_Y(X)$ then the canonical projection $X \to X/G$ is a regular smooth covering. But not all groups of homeomorphisms of X are of this form. So the question is for what groups the projection onto the orbit space has nice properties. We are interested in the case that X is a compact Riemann surface, and the orbit space is to be also a Riemann surface such that the projection is holomorphic.

Such a map is always a finitely branched covering (see [**FK92**, I.2.4]), and a necessary and sufficient condition on a group G that the canonical projection $X \to X/G$ is a finitely branched covering is that G acts *properly discontinuously* on X (see [**FK92**, IV.9]). We will not define here what this means because in the special case $X = \mathcal{U}$, this is equivalent to the condition that the orbits of G on $\mathcal{U}$ are discrete (see [**FK92**, p. 3]). Moreover, by Theorem 1.4, the group $\mathrm{Aut}(\mathcal{U})$ is isomorphic to $PSL_2(\mathbb{R})$, a factor group of $SL_2(\mathbb{R})$ which is a topological space via its natural embedding into $\mathbb{R}^4$. So $\mathrm{Aut}(\mathcal{U})$ can be endowed with the quotient topology and hence is itself a topological space. One can show that the orbits of a subgroup G of $\mathrm{Aut}(\mathcal{U})$ on $\mathcal{U}$ are discrete if and only if G is discrete as a subspace of $\mathrm{Aut}(\mathcal{U})$ (see [**JS87**, Corollary 5.6.4]), thus the following definition describes the groups whose orbit spaces on $\mathcal{U}$ are Riemann surfaces.

DEFINITION 2.3. A *Fuchsian group* is a discrete subgroup of $\mathrm{Aut}(\mathcal{U})$. A torsion-free Fuchsian group is called a *Fuchsian surface group*.

2.5. Lifts of Automorphisms.

LEMMA 2.4. *Let X be a Riemann surface with universal covering $\pi \colon \tilde{X} \to X$, and $\sigma \colon X \to X$ a (conformal) homeomorphism. Then σ can be lifted to $\tilde{X}$, i.e., there is a (conformal) homeomorphism $\tilde{\sigma} \colon \tilde{X} \to \tilde{X}$ that preserves the fibres of π, i.e., $\tilde{\sigma}\pi = \pi\sigma$.*

Proof. [**For77**, 4.7–4.18], special cases in [**JS87**, 4.18 and 5.9.3]. More general statements hold, see for example [**Acc94**, Theorem 4.11]. □

The condition $\tilde{\sigma}\pi = \pi\sigma$ means that $\tilde{\sigma}$ normalizes the group $\mathrm{Aut}_X(\tilde{X})$ of covering transformations of π. It implies that $\tilde{\sigma}$ is unique up to covering transformations, because for two lifts $\tilde{\sigma}_1$ and $\tilde{\sigma}_2$, the quotient $\tilde{\sigma}_1\tilde{\sigma}_2^{-1}$ is a covering transformation of $\tilde{X} \to X$ by

$$(\tilde{\sigma}_1\tilde{\sigma}_2^{-1})\pi = \tilde{\sigma}_1(\tilde{\sigma}_2^{-1}\pi) = \tilde{\sigma}_1\pi\sigma^{-1} = \pi\sigma\sigma^{-1} = \pi.$$

LEMMA 2.5. *For a Riemann surface X with universal covering $\tilde{X}$, we have $\mathrm{Aut}(X) \cong N_{\mathrm{Aut}(\tilde{X})}(\mathrm{Aut}_X(\tilde{X}))/\mathrm{Aut}_X(\tilde{X})$.*

Proof. So the map $\sigma \mapsto \tilde{\sigma} \cdot \mathrm{Aut}_X(\tilde{X})$ defines an injective homomorphism of $\mathrm{Aut}(X)$ into $N_{\mathrm{Aut}(\tilde{X})}(\mathrm{Aut}_X(\tilde{X}))/\mathrm{Aut}_X(\tilde{X})$. In fact it is surjective, since any $\tilde{\sigma} \in N_{\mathrm{Aut}(\tilde{X})}(\mathrm{Aut}_X(\tilde{X}))$ induces a map $\sigma \in \mathrm{Aut}(X)$ via $(x\pi)\sigma = (x\tilde{\sigma})\pi$. □

COROLLARY 2.6. *Suppose $\sigma \in \mathrm{Aut}(X)$ fixes a point $x \in X$. Then one can choose a lift $\tilde{\sigma}$ to the universal covering space $\tilde{X}$ that fixes a point in $\tilde{X}$. Moreover, all lifts of σ are conjugate in $\mathrm{Aut}(\tilde{X})$.*

Proof. (see [**Acc94**, 4.12]) Take $\tilde{x} \in \tilde{X}$ with $\tilde{x}\pi = x$, and $\tilde{\sigma}$ any lift of σ to $\tilde{X}$. Then $\tilde{\sigma}$ respects the fibres of the covering $\pi \colon \tilde{X} \to X$, we have $(\tilde{x}\tilde{\sigma})\pi = x\sigma = x = \tilde{x}\pi$. That is, $\tilde{x}$ and $\tilde{x}\tilde{\sigma}$ lie in the same fibre of π. By Theorem 2.2, there exists a covering transformation $t \in \mathrm{Aut}_X(\tilde{X})$ with $\tilde{x}(\tilde{\sigma}t) = \tilde{x}$. So $\tilde{\sigma}t$ is a lift of σ that fixes $\tilde{x}$.

Let $\tilde{\sigma}_1$ and $\tilde{\sigma}_2$ be two lifts of σ that fix the points $\tilde{x}_1$ and $\tilde{x}_2$, respectively. Again by Theorem 2.2, there is $t \in \mathrm{Aut}_X(\tilde{X})$ with $\tilde{x}_1 t = \tilde{x}_2$, and $t\tilde{\sigma}_2 t^{-1}$ fixes $\tilde{x}_1$. But then $t\tilde{\sigma}_2 t^{-1} = \tilde{\sigma}_1$ because the quotient of the elements on the two sides is a covering transformation that fixes a point, i.e., it is the identity on $\mathcal{U}$. □

REMARK 2.7. A group theoretic interpretation of the above statement about lifts of automorphisms with fixed points is that $\langle \tilde{\sigma} \rangle$ is a complement of

$\text{Aut}_X(\tilde{X})$ in the extension $\langle \text{Aut}_X(\tilde{X}), \tilde{\sigma} \rangle$, and that all such complements are conjugate.

3. Fuchsian Groups

A Riemann surface X will be parametrized as the orbit space of its universal covering space $\tilde{X}$, under the action of the subgroup $\text{Aut}_X(\tilde{X})$ of $\text{Aut}(\tilde{X})$. We are interested in compact Riemann surfaces of genus at least 2, in all these cases we have $\tilde{X} = \mathcal{U}$, the upper half plane (see [**JS87**, Theorem 4.19.8]). Before we look more closely at the groups that occur as $\text{Aut}_X(\mathcal{U})$, we collect some facts about the action of $\text{Aut}(\mathcal{U})$ on $\mathcal{U}$.

3.1. Hyperbolic Distance and Area. For a piecewise differentiable path $\gamma \colon [0,1] \to \mathcal{U}$ with $\gamma(t) = x(t) + iy(t)$, we set

$$h(\gamma) = \int_0^1 \frac{1}{y} \sqrt{\left(\frac{dx}{dt}\right)^2 + \left(\frac{dy}{dt}\right)^2}\, dt$$

and define the *hyperbolic distance* ρ on $\mathcal{U}$ by

$$\rho(z, w) = \inf_\gamma \{h(\gamma)\}$$

where the infimum is taken over all piecewise differentiable paths $\gamma \colon [0,1] \to \mathcal{U}$ with $\gamma(0) = z$ and $\gamma(1) = w$. Together with ρ, $\mathcal{U}$ is a metric space. Moreover, $\text{Aut}(\mathcal{U})$ acts as a group of *isometries* with respect to ρ.

In the definition of $\rho(z, w)$, the infimum is in fact a minimum, and there is a unique path of shortest hyperbolic length that joins z and w, which is called an *H-line segment*.

THEOREM 3.1. *The H-line segments in $\mathcal{U}$ are arcs of semi-circles with center on the real axis or segments of Euclidean lines perpendicular to the real axis.*

Proof. [**JS87**, Theorem 5.3.3] $\quad\square$

The *hyperbolic area* of a subset $E \subseteq \mathcal{U}$ is defined as

$$\mu(E) = \iint_E \frac{dx\, dy}{y^2}$$

if this integral exists. Then $\mu(E)$ is invariant under all transformations in $\text{Aut}(\mathcal{U})$.

3.2. A Presentation for Fuchsian Groups.

THEOREM 3.2. *If Γ is a Fuchsian group with compact orbit space $\mathcal{U}/\Gamma$ of genus g then there are elements $a_1, b_1, a_2, b_2, \ldots, a_g, b_g, c_1, c_2, \ldots, c_r$ in $\mathrm{Aut}(\mathcal{U})$ such that the following hold.*

1. *We have $\Gamma = \langle a_1, b_1, a_2, b_2, \ldots, a_g, b_g, c_1, c_2, \ldots, c_r \rangle$.*
2. *Defining relations for Γ are given by*

$$c_1^{m_1}, c_2^{m_2}, \ldots, c_r^{m_r}, \prod_{i=1}^{g} [a_i, b_i] \prod_{j=1}^{r} c_j,$$

   where the m_i are integers with $2 \leq m_1 \leq m_2 \leq \cdots \leq m_r$.
3. *Each nonidentity element of finite order in Γ lies in a unique conjugate of $\langle c_i \rangle$ for suitable i. Furthermore, the cyclic groups $\langle c_i \rangle$ are self-normalizing in Γ.*
4. *Each nonidentity element of finite order in Γ (a so-called* elliptic element*) has a unique fixed point in $\mathcal{U}$. Each element of infinite order in Γ (a so-called* hyperbolic *element) acts fixed point freely on $\mathcal{U}$.*

Proof. See [**Leh64**, p. 227 and p. 234] or [**Sah69**, Appendix]. □

3.3. Signatures. For a Fuchsian group Γ as in Theorem 3.2, the numbers g, r, and m_1, m_2, $\ldots$, m_r are uniquely determined; note that $2g$ is the rank of the free abelian part of the commutator factor group of Γ, see Lemma A.3. The following lemma shows that the ordering of the values m_i does not impose any extra condition.

LEMMA 3.3. *Let G be a group, and c_1, c_2, $\ldots$ $c_r \in G$. For each permutation π of $\{1, 2, \ldots, r\}$, there are elements h_1, h_2, $\ldots$, $h_r \in G$ such that $\prod_{i=1}^{r}(h_i c_{\pi(i)} h_i^{-1})$ is G-conjugate to $\prod_{i=1}^{r} c_i$.*

Proof. Without loss of generality assume that $\pi = (i, i+1)$, with $1 \leq i \leq r - 1$, and observe that $c_i c_{i+1} = (c_i c_{i+1} c_i^{-1}) c_i$. □

We call $(g; m_1, m_2, \ldots, m_r)$ the *signature* of Γ, the integer g is called its *orbit genus*, and the m_i are called the *periods* of Γ.

As a consequence of Theorem 3.2, the isomorphism type of Γ is determined by the signature of Γ.

REMARK 3.4. In the literature, more general signatures are also considered that may have infinite periods, the signatures we have introduced are then called *finite signatures*, see for example [**Sah69**, Appendix]. The orbit spaces of groups with infinite signatures are not compact, so we are only interested in Fuchsian groups with finite signatures.

3.4. Groups of Genera 0 and 1. Compact Riemann surfaces of genus 0 and 1 and finite subgroups of their –infinite– automorphism groups can also be uniformized by groups Γ with finite signatures and torsion-free normal subgroups $K \leq \Gamma$. Since the universal covering spaces of these Riemann surfaces are $\hat{\mathbb{C}}$ and $\mathbb{C}$, respectively, the groups Γ are subgroups of $\text{Aut}(\hat{\mathbb{C}})$ and $\text{Aut}(\mathbb{C})$.

The possible signatures for Γ in the case that K has orbit genus 0 satisfy $-1 + \frac{1}{2}\sum_{i=1}^{r}(1 - 1/m_i) < 0$, K is trivial, and the groups Γ are finite polyhedral groups; they are exactly the finite subgroups of the infinite group $\text{Aut}(\hat{\mathbb{C}})$. Table 1 lists the signatures and the isomorphism types of these groups. Note that the signature of a polyhedral group is uniquely determined only if the periods are forced to be element orders in the group in question; thus we exclude in the following "signatures" $(0; m)$, with arbitrary $m \geq 2$, for the trivial group and $(0; m_1, m_2)$, with $m_1 \neq m_2$, for the cyclic group of order $\gcd(m_1, m_2)$.

$(g_0; m_1, \ldots, m_r)$	Γ	$|\Gamma|$
$(0; -)$	1	1
$(0; n, n)$	n	n
$(0; 2, 2, n)$	D_{2n}	$2n$
$(0; 2, 3, 3)$	A_4	12
$(0; 2, 3, 4)$	S_4	24
$(0; 2, 3, 5)$	A_5	60

TABLE 1. Signatures and Groups in Genus 0

The possible signatures for Γ in the case that K has orbit genus 1 satisfy $-1 + \frac{1}{2}\sum_{i=1}^{r}(1 - 1/m_i) = 0$. Table 2 lists the signatures and the isomorphism types of the groups. The full automorphism group of any Riemann surface of genus 1 is infinite, and isomorphic to one of $(\mathbb{C}/\mathbb{Z}^2)\colon 2$, $(\mathbb{C}/\mathbb{Z}^2)\colon 4$, and $(\mathbb{C}/\mathbb{Z}^2)\colon 6$.

$(g_0; m_1, \ldots, m_r)$	Γ
$(1; -)$	$\mathbb{Z}^2$
$(0; 2, 3, 6)$	$\mathbb{Z}^2\colon 6$
$(0; 2, 4, 4)$	$\mathbb{Z}^2\colon 4$
$(0; 3, 3, 3)$	$\mathbb{Z}^2\colon 3$
$(0; 2, 2, 2, 2)$	$\mathbb{Z}^2\colon 2$

TABLE 2. Signatures and Groups in Genus 1

For more details, see [**Sah69**, Appendix], [**JS87**, Theorem 2.13.5] or [**FK92**, IV.9.3].

3.5. Fundamental Regions. Each Fuchsian group Γ has a *fundamental region* in $\mathcal{U}$, that is, a closed subset $F \subseteq \mathcal{U}$ with the properties that $\bigcup_{\sigma \in \Gamma} F\sigma = \mathcal{U}$ and that the intersection $F \cap F\sigma$ is contained in the boundary of F for all $\sigma \in \Gamma$. In particular, if $p \in \mathcal{U}$ is a point with trivial stabilizer in Γ then

$$D_p(\Gamma) = \{z \in \mathcal{U} \mid \rho(z,p) \le \rho(z,p\gamma) \quad \text{for all} \quad \gamma \in \Gamma\}$$

is a connected fundamental region for Γ, the so-called *Dirichlet region for Γ centered at p* (see [**JS87**, Theorem 5.8.3], for the existence of points with trivial stabilizer see [**JS87**, Theorem 5.6.3 (ii)]). If $\mathcal{U}/\Gamma$ is compact then the boundary of a Dirichlet region is a finite number of H-line segments and thus has zero hyperbolic area.

THEOREM 3.5. *If F is a fundamental region whose boundary has zero hyperbolic area for a Fuchsian group with signature $(g; m_1, m_2, \ldots, m_r)$ then we have*

$$\mu(F) = 2\pi \left((2g - 2) + \sum_{i=1}^{r} (1 - 1/m_i) \right).$$

Proof. [**JS87**, Theorem 5.10.3]   □

Let F be a Dirichlet region for Γ, and s a side of F, i.e., one of the hyperbolic arcs that form the boundary of F. If the image $s\gamma$ of s under $\gamma \in \Gamma^\times$ is also a side of F then $s\gamma$ is called the *conjugate side* of s. One can show that the boundary of F consists of conjugate pairs of sides (see [**JS87**, pp. 245–247]).

If a side s is self-conjugate then the conjugating transformation fixes the point in the middle of s. In this case, we count each half of s as a side of its own, and consider the point in the middle as a vertex of F.

The vertices of F that are points with nontrivial stabilizer in Γ are called *elliptic* vertices, the others are called *accidental*. Note that points with nontrivial stabilizer *must* be vertices of each fundamental region that contains them, whereas the position of accidental vertices depends on the fundamental region, e.g., on the choice of the center of a Dirichlet region.

It may happen that several vertices of a fundamental region lie in the same orbit of Γ, but we may choose the fundamental region such that no two elliptic vertices are congruent modulo Γ. The boundary of such a fundamental region is called a *canonical polygon*. Each pair of conjugate sides meets in an elliptic vertex of a canonical polygon.

If Γ has r conjugacy classes of maximally cyclic subgroups and orbit genus 0 then F has $2r$ sides and $2r$ vertices, r accidental vertices and r elliptic vertices, one for each conjugacy class of elliptic generators of Γ. In a walk around a canonical polygon for Γ, elliptic and accidental vertices alter. If the stabilizer of the j-th elliptic vertex e_j has order m_j then the angle between the two sides of the polygon that meet in e_j is $2\pi/m_j$, and the stabilizer of e_j is generated by a transformation $c_j \in \Gamma$ that acts as a rotation by $2\pi/m_j$ in e_j and thus has order m_j. By successive application of the c_j to the accidental vertices of F, we see that they lie in the same Γ-orbit, and that the product $c_1 c_2 \cdots c_r$ fixes the first accidental vertex. But since each accidental vertex has trivial stabilizer in Γ, this product is the identity. (If the orbit genus g of Γ is positive then analogous statements hold, for example, the number of sides of a canonical polygon is $4g + 2r$.) This way we can deduce the relators of the presentation in Theorem 3.2.

Proofs of the above statements and more details about fundamental regions can be found in [**Leh64**, Section I.2 and Chapter IV] and [**JS87**, Section 5.8].

3.6. Normal Subgroups of Fuchsian Groups. The signature of a normal subgroup of finite index in a Fuchsian group can be computed as follows.

LEMMA 3.6 (see [**Sah69**, Appendix]). *Let Γ be a Fuchsian group with signature $(g; m_1, m_2, \ldots, m_r)$, and $H \le \Gamma$ a normal subgroup of finite index d such that $c_i H$ has order t_i in the factor group Γ/H. Then the orbit genus g' of H is given by*

$$g' - 1 = d(g - 1) + \frac{d}{2} \sum_{i=1}^{r} (1 - 1/t_i),$$

and the periods of H are $f_{ij} = m_i/t_i$, $1 \le j \le d/t_i$, $1 \le i \le r$, where $f_{ij} = 1$ are deleted.

Proof. Let $y \ne 1$ be an elliptic generator of H. Then there is a unique index i such that y is conjugate to c_i^t in Γ. The order of $c_i H$ in Γ/H is t_i, so $c_i^{t_i}$ is the smallest power of c_i that is contained in H, and thus $\langle c_i^t \rangle = \langle c_i^{t_i} \rangle$, that is, y has order m_i/t_i. The statement about the periods of H is proved if we show that the Γ-class $(c_i^{t_i})^\Gamma$ of $c_i^{t_i}$ splits into d/t_i classes in H.

For that, note that the class y^Γ is the union of H-classes $(y^\gamma)^H$, where γ runs over a set of coset representatives of H in Γ. More precisely, this union of H-classes is disjoint if γ runs over a set of coset representatives of $H \cdot C_\Gamma(y)$ in Γ. But $C_\Gamma(c_i^{t_i}) = \langle c_i \rangle$, and

$$[\Gamma{:}(H \cdot \langle c_i \rangle)] = [\Gamma{:}H]/[(H \cdot \langle c_i \rangle){:}H] = d/[\langle c_i \rangle{:}(H \cap \langle c_i \rangle)] = d/t_i.$$

The orbit genus g' is computed using Theorem 3.5. Let F_Γ and F_H be fundamental regions of Γ and H, respectively. Then

$$d = \frac{\mu(F_H)}{\mu(F_\Gamma)} = \frac{2\pi(2g' - 2 + \sum_{i=1}^{r} \sum_{j=1}^{d/t_i}(1 - 1/f_{ij}))}{2\pi(2g - 2 + \sum_{i=1}^{r}(1 - 1/m_i))},$$

so

$$2g' - 2 = d(2g - 2) + \sum_{i=1}^{r} (d(1 - 1/m_i) - d/t_i(1 - t_i/m_i))$$

$$= d(2g - 2) + d \sum_{i=1}^{r} (1 - 1/t_i).$$

$\square$

EXAMPLE 3.7 (see also [**Sah69**, Theorem 1.6 (c)]). Let p be a prime, and consider the group $\Gamma = \Gamma(0; p, p, p)$. Its commutator factor group is an elementary abelian group of order p^2 by Lemma A.3, so Γ has exactly $p + 1$ normal subgroups of index p. Three of them arise as kernels of epimorphisms that map exactly one of the elliptic generators of Γ to the identity of a cyclic group of order p, and have signature $(0; p, p, \ldots, p)$, the number of periods being p. The other $p - 2$ normal subgroups of index p contain none of the elliptic generators and hence are torsion-free groups, with signature $(\frac{p-1}{2}; -)$. In particular, the derived subgroup of Γ is torsion-free, and it has orbit genus $\frac{1}{2}(p - 1)(p - 2)$.

Note that Lemma 3.6 is formulated only for Fuchsian groups Γ, that is, for $p \geq 5$. But the statements of the lemma (and hence of this example) hold also for $p = 2$ and $p = 3$, where the signature $(0; p, p, p)$ corresponds to the genera 0 and 1, respectively. $\diamond$

EXAMPLE 3.8. Let $\Gamma = \Gamma(0; 2, 3, 8)$. By Lemma A.3, the derived subgroup Γ' is of index 2 in Γ, its signature is $(0; 3, 3, 4)$ by Lemma 3.6. With the same argument, the second derived subgroup Γ'' has index 3 in Γ', and signature $(0; 4, 4, 4)$. Since Γ has no normal subgroup of index 3, we have $\Gamma/\Gamma'' \cong S_3$, the unique nonabelian group of order 6. Again by Lemma 3.6, the three conjugate subgroups of index 3 in Γ have signature $(0; 2, 4, 8)$; note that only divisors of 3 and 8 can occur as periods of subgroups of Γ.

The commutator factor group of Γ'' is of type 4×4. Each of the three subgroups of index 2 in Γ'' contains exactly one of the three elliptic generators of order 4, so the groups all have the signature $(0; 2, 2, 4, 4)$, in fact they are conjugate in Γ', and their intersection has signature $(0; 2, 2, 2, 2, 2, 2)$. In this group, we find a group with signature $(2; -)$ of index 2.

Thus we have found an epimorphism from Γ onto a group of order 48, with torsion-free kernel. The factor group can be shown to be isomorphic with $GL_2(3)$. In other words, we have established the group $GL_2(3)$ and all its subgroups as groups of automorphisms of a compact Riemann surface of genus 2. $\diamond$

3.7. Uniformization by Fuchsian Surface Groups.

THEOREM 3.9. *Any compact Riemann surface X of genus $g \geq 2$ is conformally equivalent to the orbit space $\mathcal{U}/K$ for a Fuchsian surface group K with orbit genus g. Furthermore, $\mathcal{U}$ is the universal covering of X, and $K = \mathrm{Aut}_X(\mathcal{U})$ is unique up to conjugacy in $\mathrm{Aut}(\mathcal{U})$.*

Proof. [**JS87**, 4.19.8 and 5.9.3]   □

If the Riemann surface X is conformally equivalent to the orbit space $\mathcal{U}/K$ for $K \leq \mathrm{Aut}(\mathcal{U})$ then we say that X is *uniformized* by K.

Note that the orbit spaces $\mathcal{U}/\Gamma$ with Γ a Fuchsian group as in Theorem 3.2 are also compact Riemann surfaces, with genus equal to the orbit genus of Γ, if Γ does not act fixed point freely on $\mathcal{U}$. Theorem 3.9 allows us to restrict our attention to torsion-free Fuchsian groups, and it states a bijection of the conjugacy classes of these groups in $\mathrm{Aut}(\mathcal{U})$ and the compact Riemann surfaces of genus at least 2, up to conformal equivalence.

We are interested in automorphism groups of Riemann surfaces. So let X be uniformized by the Fuchsian surface group $K = \mathrm{Aut}_X(\mathcal{U}) \leq \mathrm{Aut}(\mathcal{U})$. Each element of $\mathrm{Aut}(X)$ can be lifted to an element of $\mathrm{Aut}(\mathcal{U})$ by Lemma 2.4, and the lifts of any subgroup H of $\mathrm{Aut}(X)$ together with K generate a subgroup Γ of $\mathrm{Aut}(\mathcal{U})$ that normalizes K. By these facts, we can parametrize the pairs (X, H) of Riemann surfaces X and subgroups $H \leq \mathrm{Aut}(X)$ by the pairs (K, Γ) of Fuchsian surface groups K and subgroups $\Gamma \leq N_{\mathrm{Aut}(\mathcal{U})}(K)$. The next theorem shows that the normalizing groups that occur are Fuchsian.

THEOREM 3.10. *Let K be a noncyclic Fuchsian group. Then the normalizer of K in $\mathrm{Aut}(\mathcal{U})$ is a Fuchsian group.*

Proof. [**JS87**, 5.7.5]   □

Now we change our perspective, and regard a pair (K, Γ) as a pair of a Fuchsian group Γ and a torsion-free normal subgroup K. From this viewpoint, the question for Riemann surfaces of a given genus and with a prescribed (abstract) group of automorphisms can be answered as follows.

THEOREM 3.11. *A finite group G acts as a group of automorphisms of some compact Riemann surface of genus $g \geq 2$, if and only if G is isomorphic to Γ/K where Γ is a Fuchsian group with compact orbit space, and K is a Fuchsian surface group with orbit genus g that is a normal subgroup of Γ.*

Proof. [**JS87**, 5.9.5]   □

This looks already much like a translation of the problem of finding certain Riemann surfaces into the group theoretic problem of finding certain normal subgroups in a given group. The following theorem justifies this approach.

THEOREM 3.12 (Poincaré). *The vector $(g; m_1, m_2, \ldots, m_r)$ is the signature of a Fuchsian group Γ with compact orbit space $\mathcal{U}/\Gamma$ of genus g if and only if*

$$(g - 1) + \frac{1}{2} \sum_{i=1}^{r} (1 - 1/m_i) > 0.$$

Proof. The proof is sketched in [**JS87**, 5.10.5], see also [**Leh64**, Note 17a on p. 403]. For a rigorous proof, see [**Mas71**]. □

With the help of Theorem 3.12, we can take an abstract group Γ with presentation as in Theorem 3.2 that satisfies the condition on the signature, specify a torsion-free normal subgroup K of finite index, and then interpret the factor group Γ/K as automorphism group of a Riemann surface, namely the Riemann surface that is uniformized by the image of K under a suitable embedding of Γ into $\text{Aut}(\mathcal{U})$.

3.8. Surface Kernel Homomorphisms. Suppose Γ is a Fuchsian group with signature as in Theorem 3.2, and that the Riemann surface X is uniformized by a normal subgroup K of finite index, with $\Gamma/K \cong G$. Then the genera of X and X/G, i.e., the orbit genera of K and Γ, are related by the Riemann–Hurwitz Formula below. It describes just the special case of Lemma 3.6 that the order of $c_i H$ is m_i for $1 \le i \le r$.

LEMMA 3.13 (Riemann–Hurwitz Formula for orbit spaces).

$$g(X) - 1 = |G| \left(g(X/G) - 1 \right) + \frac{|G|}{2} \sum_{i=1}^{r} (1 - 1/m_i).$$

This means for example that for given groups G and Γ, the genus of possible Riemann surfaces that are uniformized by normal subgroups of Γ with factor group G is fixed. As discussed above, the existence of such Riemann surfaces is equivalent to the existence of epimorphisms $\Gamma \to G$ with torsion-free kernels.

Whenever we establish the existence of certain Riemann surfaces, this will be done more or less by writing down a suitable epimorphism. As an immediate consequence of Lemma 3.6, we have

THEOREM 3.14 (see [**Har66**, Theorem 3]). *Let Φ be a homomorphism from a Fuchsian group Γ with presentation as in Theorem 3.2 onto a finite group. Then the kernel of Φ is torsion-free if and only if Φ preserves the orders of the generators $c_1, c_2, \ldots, c_r$.*

Such a homomorphism Φ with torsion-free kernel is called a *surface kernel homomorphism*.

COROLLARY 3.15 (see [**Hur93**]). *Any finite group is isomorphic to a group of automorphisms of some compact Riemann surface (of genus at least 2).*

Proof. Let G be a finite group generated by elements $\sigma_1, \sigma_2, \ldots, \sigma_s$, with $s > 1$. Then G is an epimorphic image of $\Gamma(s; -)$, via the homomorphism Φ defined by $\Phi(a_i) = \sigma_i$, $\Phi(b_i) = 1$. This clearly has torsion-free kernel because $\Gamma(s; -)$ itself is torsion-free.   □

EXAMPLE 3.16 (see [**Har71**, Lemma 6]). Let $\langle h \rangle$ be a cyclic group of order m, and Γ as in Theorem 3.2. Define $\Phi(c_i) = h^{k_i}$ for $1 \leq i \leq r$, and choose h as image of all a_i and b_i. Then Φ extends to a group homomorphism $\Gamma \to \langle h \rangle$ if and only if $\prod_{i=1}^{r} c_i$ and all $c_i^{m_i}$ are mapped to the identity in $\langle h \rangle$. This is equivalent to $\sum_{i=1}^{r} k_i \equiv 0 \pmod{m}$ and $m \mid m_i k_i$ for $1 \leq i \leq r$. Φ is a surface kernel homomorphism if additionally the orders of c_i and h^{k_i} are the same, that is, $m_i = m/\gcd(m, k_i)$ for $1 \leq i \leq r$.   ◊

Besides the technical importance of surface kernel epimorphisms for establishing the existence of certain Riemann surfaces, it is the epimorphism that relates the factor group of Γ acting as automorphism group to the given abstract group G, and thus allows us to talk about the action of elements in G on the Riemann surface. We will state this precisely in Section 3.11.

3.9. Hurwitz's Bound. Now we show that automorphism groups of compact Riemann surfaces of genus at least 2 are finite.

THEOREM 3.17 (Hurwitz). *Let X be a compact Riemann surface of genus $g \geq 2$. Then $|\mathrm{Aut}(X)| \leq 84(g - 1)$. If this bound is attained then X is uniformized by $K \leq \mathrm{Aut}(\mathcal{U})$ with $N_{\mathrm{Aut}(\mathcal{U})}(K) \cong \Gamma(0; 2, 3, 7)$, in particular $\mathrm{Aut}(X)$ is an epimorphic image of the group $\Gamma(0; 2, 3, 7)$.*

Proof. (See [**JS87**, Theorem 5.11.1] or [**FK92**, p.260].) Let $\Gamma = N_{\mathrm{Aut}(\mathcal{U})}(K)$, and choose fundamental regions F_Γ, F_K of Γ and K, respectively. Then

$$|\mathrm{Aut}(X)| = |\Gamma/K| = \frac{\mu(F_K)}{\mu(F_\Gamma)} = \frac{2\pi(2g - 2)}{\mu(F_\Gamma)}.$$

Now [**JS87**, Theorem 5.10.7] yields that $\mu(F_\Gamma) \geq \pi/21$, with equality if and only if $\Gamma \cong \Gamma(0; 2, 3, 7)$; the proof of this theorem can be carried out purely combinatorially using only Lemma 3.13.

Alternatively, one could prove by other means that $\mathrm{Aut}(X)$ is finite for any compact Riemann surface X of genus at least 2; once this is shown, the upper bound on the order of $\mathrm{Aut}(X)$ can be proved by the same combinatorial argument. For this variant, see for example [**For77, FK92, Mir95**]. (Similar elementary combinatorics will be used in the proofs of Lemma 3.18 and Corollary 9.6.)   □

Better bounds can be derived easily, for example as follows.

LEMMA 3.18. *Let G be a finite epimorphic image of the group Γ with signature $(g_0; m_1, m_2, \ldots, m_r)$, such that the kernel is torsion-free of orbit genus $g \geq 2$. Then we have the following.*

(a) *If $g_0 > 0$ or $r \geq 5$ then $|G| \leq 4(g - 1)$.*
(b) *If $r = 4$ then $|G| \leq 12(g - 1)$.*
(c) *If $|G| \geq 24(g - 1)$ then $g_0 = 0$, $r = 3$, and one of the cases listed in Table 3 occurs.*

$(g_0; m_1, m_2, m_3)$	$|G|$
$(0; 2, 3, 7)$	$84(g - 1)$
$(0; 2, 3, 8)$	$48(g - 1)$
$(0; 2, 4, 5)$	$40(g - 1)$
$(0; 2, 3, 9)$	$36(g - 1)$
$(0; 2, 3, 10)$	$30(g - 1)$
$(0; 2, 3, 11)$	$(132/5)(g - 1)$
$(0; 2, 3, 12)$	$24(g - 1)$
$(0; 2, 4, 6)$	$24(g - 1)$
$(0; 3, 3, 4)$	$24(g - 1)$

TABLE 3. Signatures for "Large" Groups of Automorphisms

Proof. By the Riemann–Hurwitz Formula of Lemma 3.13,

$$(*) \qquad g - 1 = |G|(g_0 - 1) + \frac{|G|}{2} \sum_{i=1}^{r} (1 - 1/m_i).$$

If $g_0 \geq 2$ then $g - 1 \geq |G|$. If $g_0 = 1$ then $g - 1 = (|G|/2) \cdot \sum_{i=1}^{r}(1 - 1/m_i)$, so $r \neq 0$ and hence $g - 1 \geq |G|/4$.

Now assume $g_0 = 0$. If $r \geq 5$ then $g - 1 \geq -|G| + (|G|/2) \cdot 5 \cdot 1/2 = |G|/4$. If $r = 4$ then at least one of the m_i must be larger than 2, thus $g - 1 \geq -|G| + (|G|/2) \cdot (3 \cdot 1/2 + 2/3) = |G|/12$.

Now assume $|G| \geq 24(g - 1)$. Clearly $g_0 = 0$, $r = 3$, and $2 \leq m_1 \leq m_2 \leq m_3$. If $m_1 \geq 3$ then $m_3 > 3$, because otherwise the right hand side of equation $(*)$ would be zero, and it is minimal for $(m_1, m_2, m_3) = (3, 3, 4)$, with $|G| = 24(g - 1)$. So all other possibilities must satisfy $m_1 = 2$, and then $m_2 \geq 3$ because the right hand side of $(*)$ must be nonnegative.

If $m_1 = 2$ and $m_2 = 3$ then only $7 \leq m_3 \leq 12$ satisfy the requirement on $|G|$. If $m_1 = 2$ and $m_2 = 4$ then we must have $5 \leq m_3 \leq 6$. If $m_1 = 2$ and $m_2 \geq 5$ then also $m_3 \geq 5$, hence $g - 1 \geq (|G|/2) \cdot (-2 + 1/2 + 2 \cdot 4/5) = |G|/20$, contrary to our assumption. $\qquad\square$

3.10. Rotation Constants. In this section, it will be convenient to choose the unit disk $\mathcal{D}$ instead of $\mathcal{U}$ as universal covering space of Riemann surfaces (see Example 1.3).

LEMMA 3.19 (cf. [**Gue82**, pp. 217 f.], [**Mir95**, Corollary III.3.5]). *Let X be a compact Riemann surface of genus at least 2, and $\sigma \in \mathrm{Aut}(X)$ of order $m > 1$ such that $x^\sigma = x$ for a point $x \in X$.*

There is a unique primitive complex m-th root of unity ζ such that any lift $\tilde{\sigma}$ of σ to $\mathcal{D}$ that fixes a point in $\mathcal{D}$ (see Corollary 2.6) is conjugate to the transformation $z \mapsto \zeta \cdot z$ in $\mathrm{Aut}(\mathcal{D})$.

We write $\zeta_x(\sigma) = \zeta$ and call ζ^{-1} the rotation constant of σ in x.

Proof. Let $\tilde{\sigma} \in \mathrm{Aut}(\mathcal{U})$ be a lift that fixes $\tilde{z} \in \mathcal{D}$, say. $\mathrm{Aut}(\mathcal{D})$ acts transitively on $\mathcal{D}$, so there is an element $t \in \mathrm{Aut}(\mathcal{D})$ with $\tilde{z}t = 0$, so $t^{-1}\tilde{\sigma}t$ is a conjugate of $\tilde{\sigma}$ that fixes 0.

Schwarz's Lemma (see [**JS87**, p. 201]) asserts that the stabilizer of 0 in $\mathrm{Aut}(\mathcal{D})$ is exactly the set $\{z \mapsto e^{i\vartheta} \cdot z \mid \vartheta \in \mathbb{R}\}$, and two transformations $z \mapsto \lambda_1 \cdot z$, $z \mapsto \lambda_2 \cdot z$ are conjugate in $PSL_2(\mathbb{C})$ if and only if $\lambda_1 = \lambda_2^{\pm 1}$, see [**JS87**, Theorem 2.9.3]. So we have to show that $z \mapsto \lambda \cdot z$ and $z \mapsto \lambda^{-1} \cdot z$ are *not* conjugate in $\mathrm{Aut}(\mathcal{D})$ for $\lambda \neq \pm 1$.

For that, assume $\lambda \neq \pm 1$, and suppose that $\tau = (z \mapsto (az + b)/(cz + d)) \in \mathrm{Aut}(\mathcal{D})$, with $ad - bc = 1$, conjugates $z \mapsto \lambda^{\pm 1} \cdot z$, i.e., $\tau \circ (z \mapsto \lambda \cdot z) = (z \mapsto \lambda^{-1} \cdot z) \circ \tau$, which is equivalent to $(\lambda az + b)/(\lambda cz + d) = \lambda^{-1}(az + b)/(cz + d)$ for all $z \in \mathcal{D}$. Evaluating both sides at $z = 0$ yields $b/d = \lambda^{-1}b/d$ and hence $b = 0$; note that $d \neq 0$ because $\tau \in \mathrm{Aut}(\mathcal{D})$. Multiplication by the denominators gives $\lambda az(cz + d) = \lambda^{-1}az(\lambda cz + d)$, thus $ac = ad = 0$, contrary to the assumption that $ad - bc = 1$.

Now $\tilde{\sigma}^m$ is a lift of the identity map on X, hence a covering transformation of $\mathcal{D} \to X$ that fixes a point, so $\tilde{\sigma}$ is an element of order m. It is conjugate to a unique element $z \mapsto \lambda \cdot z$ in $\mathrm{Aut}(\mathcal{D})$, thus λ is a primitive m-th root of unity in $\mathbb{C}$.

(Note that λ does not depend on the choice of the lift $\tilde{\sigma}$, by Corollary 2.6.) $\qquad\square$

REMARK 3.20. Alternatively, we could have defined rotation constants analytically. That is, if $\mathcal{A}$ is an atlas for the analytic structure on X then there is a local coordinate φ that is compatible with $\mathcal{A}$, defined in a neighborhood of the fixed point x of σ, and vanishing in x such that σ is described by $z \mapsto \zeta_x(\sigma) \cdot z$ with respect to φ. Of course existence and uniqueness of $\zeta_x(\sigma)$ must be shown then.

From this viewpoint, we get the connection between the actions on X and its universal covering $\pi \colon \mathcal{D} \to X$ as follows. The projection π is a holomorphic map, and we can choose a neighborhood of the fixed point $\tilde{z}$ of $\tilde{\sigma}$ in $\mathcal{D}$ such that $\varphi \circ \pi$ is a local coordinate on $\mathcal{D}$. Via this map, the action of $\tilde{\sigma}$ near $\tilde{z}$ is described by $z \mapsto \zeta_x(\sigma)$. In other words, $\tilde{\sigma}$ has the same rotation constant in $\tilde{z}$ as σ has in x.

The important point is that by the discussion in Section 3.5, we can extend Theorem 3.2.

THEOREM 3.21. *In Theorem 3.2, the elliptic generators of Γ can be chosen as follows.*

   5. *For $1 \le j \le r$, we have $\zeta_{z_j}(c_j) = \zeta_{m_j}$, where $\zeta_m = e^{2\pi i/m}$. That is, c_j is described by $z \mapsto \zeta_{m_j} \cdot z_j$ near its fixed point z_j.*

More about rotation constants can be found in [**Har71, Gue82, Kur87**].

3.11. The Standard Situation. Finally, we describe a "standard situation" in the sense that many later statements will refer to it.

SITUATION 3.22.

- Γ is a Fuchsian group with generators as in Theorem 3.2 and Theorem 3.21, and with orbit genus g_0.
- K is a torsion-free normal subgroup of Γ, and $X = \mathcal{U}/K$ its associated compact Riemann surface of genus $g \ge 2$.
- $G \le \mathrm{Aut}(X)$ is the group of automorphisms of X induced by Γ/K. Then $G \cong \Gamma/K$, and the orbit space X/G can be identified with $(\mathcal{U}/K)/(\Gamma/K) \cong \mathcal{U}/\Gamma$.
- $\Phi \colon \Gamma \to G$ is a surface kernel epimorphism that identifies the actions of Γ and G on X. More precisely, for $\gamma \in \Gamma$, the image of $zK \in \mathcal{U}/K$ under the element $\Phi(\gamma) \in G$ is $(zK)\Phi(\gamma) = (z\gamma)K$.
- For $1 \le i \le r$, the fixed point of the generator c_i is $z_i \in \mathcal{U}$.

4. The Riemann–Hurwitz Formula

In this section, we reinterpret the Riemann–Hurwitz Formula in terms of branched coverings and fixed points.

In Situation 3.22, the canonical projection $\pi\colon \mathcal{U}/K \to \mathcal{U}/\Gamma$ is a branched covering, branched exactly over the set

$$\Delta(\pi) = \{P\pi \mid P \in \mathcal{U}/K, |\mathrm{Stab}_G(P)| > 1\} \subseteq \mathcal{U}/\Gamma,$$

where $\mathrm{Stab}_G(x) = \{\sigma \in G \mid x\sigma = x\}$ denotes the *point stabilizer* (or *isotropy subgroup*) of x in G.

Lemma 4.1. *In Situation 3.22,* $\Delta(\pi) = \{z_i\Gamma \mid 1 \le i \le r\}$.

Proof. The stabilizer of z_iK in Γ/K is nontrivial since it contains $\langle c_iK\rangle$, this means $z_i\Gamma = (z_iK)\pi \in \Delta(\pi)$.

Conversely, let zK be a point in $\mathcal{U}/K$ with $(zK)\pi \in \Delta(\pi)$, and let σ be a nonidentity element in $\mathrm{Stab}_G(zK)$. Following Corollary 2.6, we choose a lift $\tilde{\sigma}$ of σ that fixes a point $z \in \mathcal{U}$, then $\tilde{\sigma}$ is elliptic and thus conjugate to a power c_i^m of one of the elliptic generators of Γ. Let $\tilde{\sigma} = tc_i^m t^{-1}$ for $t \in \Gamma$, then $(zt)c_i^m = z(\tilde{\sigma}t) = zt$, so $zt = z_i$ and hence $(zK)\pi = z\Gamma = z_i\Gamma$. $\square$

Lemma 4.2. *Let X be a compact Riemann surface with $g(X) \ge 2$, and $G \le \mathrm{Aut}(X)$. For $x \in X$, $\mathrm{Stab}_G(x)$ is cyclic, and it is nontrivial only for finitely many points x.*

Proof. For the finitely many elements σ of $\mathrm{Stab}_G(x)$, one can choose lifts $\tilde{\sigma} \in \mathrm{Aut}(\mathcal{U})$ that fix a common point $z \in \mathcal{U}$. They generate a Fuchsian group with common fixed point z, so by [**JS87**, 5.7.2] it is cyclic.

For each point $x \in X$ with nontrivial stabilizer, the image under the natural projection $\pi\colon X \to X/G$ is one of the finitely many branch points. $\square$

Corollary 4.3. *In Situation 3.22,* $\mathrm{Stab}_G(z_iK) = \langle c_iK\rangle$.

Proof. By Corollary 2.6 and Lemma 4.2, the full preimage of $\mathrm{Stab}_G(z_iK)$ in Γ is $\langle K, \sigma\rangle$ where $\langle\sigma\rangle$ is a finite cyclic subgroup of Γ. According to Theorem 3.2, we have $\langle\sigma\rangle = \langle c_i\rangle$. $\square$

Now we can reformulate the Riemann–Hurwitz Formula of Lemma 3.13 in terms of stabilizers.

Lemma 4.4. *In Situation 3.22, we have*

$$g(X) - 1 = |G|\Big(g(X/G) - 1\Big) + \frac{1}{2}\sum_{x\in X}(|\mathrm{Stab}_G(x)| - 1).$$

Proof. For $x \in X$, exactly $|G|/|\text{Stab}_G(x)|$ points lie in the fibre of $x\pi$, so

$$\sum_{x \in X}(|\text{Stab}_G(x)| - 1) = \sum_{\substack{y \in X/G}} \sum_{\substack{x \in X \\ x\pi = y}}(|\text{Stab}_G(x)| - 1)$$

$$= |G| \sum_{x\pi \in X/G}(1 - 1/|\text{Stab}_G(x)|)$$

$$= |G| \sum_{x\pi \in \Delta(\pi)}(1 - 1/|\text{Stab}_G(x)|)$$

$$= |G| \sum_{i=1}^{r}(1 - 1/|\text{Stab}_G(z_i K)|)$$

$$= |G| \sum_{i=1}^{r}(1 - 1/m_i),$$

and the claim follows from Lemma 3.13.   □

REMARK 4.5. The formulations of the Riemann–Hurwitz Formula in Lemmas 3.6, 3.13, and 4.4 apply only to the special case of orbit spaces. For the general situation of branched coverings, see for example [**FK92**, p. 21 or p. 77]

5. Examples

5.1. Full Automorphism Groups. The nontrivial finite factor groups of $\Gamma(0; 2, 3, 7)$ are called *Hurwitz groups*, they occur as automorphism groups of maximal order of Riemann surfaces, in particular they are the full automorphism groups of these surfaces.

The finite surface kernel factors of groups with the other signatures in Table 3 than $(0; 3, 3, 4)$ occur also only as full automorphism groups because they are maximal with respect to inclusion; $\Gamma(0; 2, 3, 8)$ contains a subgroup of index 2 with signature $(0; 3, 3, 4)$.

Greenberg (see [**Gre63**]) defines a Fuchsian group Γ to be *finitely maximal* if there is no other Fuchsian group G such that $\Gamma < G$ and $[G{:}\Gamma]$ is finite. He also states that a Fuchsian group with signature $(g; m_1, m_2, \ldots, m_r)$ is finitely maximal if and only if no other Fuchsian group contains it, regardless of the finiteness of the index.

THEOREM 5.1 ([**Gre63**, Theorems 3A and 3B]). *No Fuchsian group with signature $(g; m_1, m_2, \ldots, m_r)$ is finitely maximal if and only if the signature is one of*

$$(0; 2, n, 2n), (0; 3, n, 3n), (0; m, m, n), (0; m, m, n, n), (1; 2, 2), (2; -).$$

Fuchsian groups with other signatures may or may not be finitely maximal, and at least one conjugacy class of such groups is finitely maximal. [**Gre63**, Theorem 2] states that 'most such Fuchsian groups are finitely maximal'. Two further interesting consequences are that the full automorphism groups of most compact Riemann surfaces of genus $g > 2$ are trivial, and that every finite group is isomorphic to the full automorphism group of a compact Riemann surface.

5.2. Counting Surfaces with Given Automorphism Group.

It is a well-known fact that the set of conjugacy classes of groups with signature $(g; m_1, m_2, \ldots, m_r)$ in $\mathrm{Aut}(\mathcal{U})$ is homeomorphic to $\mathbb{R}^{6g-6+2r}$, see [**Ahl54**]. So there is a unique conjugacy class of groups with signature $(g; m_1, m_2, \ldots, m_r)$ if (g, r) is either $(0, 3)$ or $(1, 0)$, in all other cases the number of these conjugacy classes is uncountable.

Every Riemann surface with maximal group of automorphisms is uniformized by a torsion-free normal subgroup of $\Gamma(0; 2, 3, 7)$. There is a unique conjugacy class of groups $\Gamma(0; 2, 3, 7)$ in $\mathrm{Aut}(\mathcal{U})$, and by Theorem 5.1, $\Gamma(0; 2, 3, 7)$ is self-normalizing in $\mathrm{Aut}(\mathcal{U})$, and thus no two normal subgroups of $\Gamma(0; 2, 3, 7)$ are conjugate in $\mathrm{Aut}(\mathcal{U})$. This means that the Riemann surfaces with maximal automorphism group up to conformal equivalence are parametrized by the torsion-free normal subgroups of $\Gamma(0; 2, 3, 7)$.

So it makes sense to count the nonisomorphic Riemann surfaces of a given genus that are uniformized by normal subgroups of $\Gamma(0; 2, 3, 7)$. For example, we will see later that $\Gamma(0; 2, 3, 7)$ has a unique normal subgroup of index 168, so there is a unique Riemann surface of genus $g = 3$ with $84(g - 1) = 168$ automorphisms (see Example 15.4), and there are exactly three Riemann surfaces of genus 14 with automorphism group $L_2(13)$ of order $84 \cdot 13 = 1\,092$ (see Example 15.31).

For factor groups of Γ that have a faithful permutation representation on few points, one can compute the normal subgroups directly by enumerating the classes of subgroups of low index in Γ. An alternative approach that will be described in Chapter 4 is based on the fact that $\mathrm{Aut}(G)$ acts on the set $\mathrm{Epi}(\Gamma, G)$ of epimorphisms $\Gamma \to G$ via composition, and that the orbits are in bijection with those normal subgroups N of Γ that satisfy $\Gamma/N \cong G$.

CHAPTER 2

Group Characters

This chapter introduces in Section 6 the notation of representation theory that is used later on, and states some well-known facts without proofs. For their validity, see for example [**Isa76**]. Then in Section 7 the module is defined whose character will play the main role afterwards. Finally, in Section 8 is discussed what it means that a character comes from a Riemann surface.

6. Basic Facts about Characters

Given a finite group G and a complex vector space V of finite dimension n, a group homomorphism ρ that maps G to a subgroup of $GL(V)$, the group of invertible $\mathbb{C}$-linear transformations of V, is called a complex *representation* of G, and V together with this action is called a G-*module*.

With respect to a basis of V, the action of any $\sigma \in G$ via ρ can be described by an invertible complex $n \times n$ matrix $\hat{\rho}(\sigma)$ such that $\hat{\rho}$ is a representation of G with image in $GL_n(\mathbb{C}) = GL(\mathbb{C}^n)$, and acting on the row space $\mathbb{C}^n$. We identify the representations ρ and $\hat{\rho}$, and also the G-modules V and $\mathbb{C}^n$.

Two matrix representations $\hat{\rho}_1, \hat{\rho}_2$ are regarded as *equivalent* if one arises from the other by a basis change, i.e., if there is a matrix $M \in GL_n(\mathbb{C})$ such that $\hat{\rho}_2(\sigma) = M^{-1}\hat{\rho}_1(\sigma)M$ for all $\sigma \in G$. In this case we also call the modules for $\hat{\rho}_1$ and $\hat{\rho}_2$ *equivalent* (or *isomorphic*).

The trace $\mathrm{Tr}(\hat{\rho}(\sigma))$ of the matrix $\hat{\rho}(\sigma)$ is invariant under conjugation, so equivalent representations have equal traces. One can show that the converse also holds, i.e., the equivalence classes of complex G-representations are characterized by their traces. The map $\chi_\rho \colon G \to \mathbb{C}$, defined by $\chi_\rho(\sigma) = \mathrm{Tr}(\rho(\sigma))$, is called the *character* afforded by ρ, or simply a G-character. The *degree* of χ_ρ is $\chi_\rho(1)$, and the *kernel* of χ_ρ is $\ker(\chi_\rho) = \{\sigma \in G \mid \chi_\rho(\sigma) = \chi_\rho(1)\}$. The kernel of χ_ρ is equal to the kernel of ρ. A G-character with trivial kernel is called *faithful*; the underlying action of G is then also called faithful (or *effective*).

We define addition of characters pointwise. The sum of two G-characters is again a character, since it is afforded by the representation that maps each group element to the block diagonal matrix given by the representations affording the summands. A character is called *irreducible* if it cannot be written as the sum of two characters. Thus every subgroup of $GL_n(\mathbb{C})$ is conjugate to a group of block diagonal matrices where the projection onto each block defines a representation with irreducible character.

The set $\mathrm{Irr}(G)$ of irreducible characters of G is finite, its cardinality is equal to the number of conjugacy classes of G. So $\mathrm{Irr}(G)$ can be given by a square matrix whose rows and columns are indexed by the irreducible characters and the conjugacy classes of G, respectively. This matrix (usually together with additional information about element orders and power maps) is called the *character table* of G.

A map $G \to \mathbb{C}$ that is constant on the conjugacy classes of G is called a *class function* of G. So characters are class functions. Other examples are differences of characters, which are called *virtual* characters.

With respect to the skew-symmetric product (or *scalar product*) defined by

$$[\chi, \psi] = \frac{1}{|G|} \sum_{\sigma \in G} \chi(\sigma)\psi(\sigma^{-1}),$$

$\mathrm{Irr}(G)$ is an orthogonal basis of the complex vector space of G-class functions.

For a subgroup H of G, the *restricted class function* of a G-class function χ of G to H is defined as $\chi_H \colon H \to \mathbb{C}$ with $\chi_H(h) = \chi(h)$ for $h \in H$, and the *induced class function* of an H-class function φ of H to G is defined as $\varphi^G \colon G \to \mathbb{C}$ with

$$\varphi^G(\sigma) = \frac{1}{|H|} \sum_{\substack{x \in G \\ x\sigma x^{-1} \in H}} \varphi(x\sigma x^{-1}).$$

Induction is transitive, i.e., for subgroups $H \leq K \leq G$ and an H-class function φ, we have $(\varphi^K)^G = \varphi^G$. Restriction and induction map characters to characters. If χ is a class function of G and φ a class function of $H \leq G$ then $[\varphi^G, \chi] = [\varphi, \chi_H]$, see [**Isa76**, Lemma 5.2]; this relation is called Frobenius reciprocity.

We denote by 1_G the *trivial character* of G, defined by $1_G(\sigma) = 1$ for all $\sigma \in G$. Clearly $1_G \in \mathrm{Irr}(G)$, and the scalar product $[\chi, 1_G]$ is the dimension of the space of fixed points under G in any G-module that affords the character χ.

Let Ω be a finite G-set, and take a complex vector space V with basis $B = (v_\omega | \omega \in \Omega)$. The action of G on B via $\sigma(v_\omega) = v_{\sigma(\omega)}$ extends to a linear action on V, and mapping each element of G to the matrix of its action with respect to B defines a representation of G. The character afforded by such a representation is called a *permutation character*.

Corresponding to the decomposition of Ω as a disjoint union of transitive G-sets, the permutation character of the action on Ω decomposes as a sum of *transitive permutation characters*. Moreover, if the subgroup H is the point stabilizer of a transitive G-set then the transitive permutation character of the action equals the induced character 1_H^G. So the character values are given by

$$1_H^G(\sigma) = \frac{|G|}{|H|} \cdot \frac{|\sigma^G \cap H|}{|\sigma^G|}.$$

Note that the right hand side of this equation is invariant under conjugation in G, so $1_H^G = 1_K^G$ for two G-conjugate subgroups H, K of G.

If H and K are subgroups of G and φ is a class function of H then $(\varphi^G)_K$ can be decomposed as follows, by a theorem of Mackey.

$$(\varphi^G)_K = \sum_{t \in T} (\varphi_{H^t \cap K}^t)^K,$$

where T is a set of double coset representatives such that $G = \biguplus_{t \in T} HtK$, and $\varphi^t(h^t) = \varphi(h)$ for all $h \in H$. Setting $\varphi = 1_H$, we see that the decomposition of $(1_H^G)_K$ into transitive permutation characters of K is given by

$$(1_H^G)_K = \sum_{t \in T} 1_{H^t \cap K}^K.$$

As a special case, we take H to be the trivial subgroup of G; then $\rho_G = 1_H^G$ is called the *regular character* of G. So ρ_G has degree $|G|$, and $\rho_G(\sigma) = 0$ for $\sigma \in G^\times$. We have $\rho_G = \sum_{\chi \in \mathrm{Irr}(G)} [\rho_G, \chi] \cdot \chi = \sum_{\chi \in \mathrm{Irr}(G)} \chi(1) \cdot \chi$.

Characters of degree 1 are called *linear*. The value of a linear character on an element of order n is an n-th complex root of unity.

If G is an abelian group then every irreducible character of G is linear, because ρ_G has $|G|$ irreducible constituents and the degree of ρ_G is $|G|$. So for every matrix representation of an abelian group, the representing matrices can be simultaneously conjugated to a group of diagonal matrices.

Applying this to the special case of cyclic groups, we see that character values for elements of order n are sums of n-th complex roots of unity, and that $\chi(\sigma^{-1}) = \overline{\chi(\sigma)}$, the complex conjugate of $\chi(\sigma)$.

More generally, if G has exponent n then consider the group $\mathrm{Aut}(\mathbb{Q}(\zeta_n))$ of field automorphisms of the n-th cyclotomic field $\mathbb{Q}(\zeta_n)$; this group consists of all elements $*k$ with k a prime residue modulo n, and $*k$ is defined by $\zeta_n^{*k} = \zeta_n^k$. For a G-character χ and k coprime to n, we may define the class function χ^{*k} by $\chi^{*k}(\sigma) = \chi(\sigma^k)$. Then χ^{*k} is also a G-character because the composition of a matrix representation with a field automorphism is again a representation. We call the characters χ^{*k} the *Galois conjugates* of χ.

So $\mathrm{Aut}(\mathbb{Q}(\zeta_n))$ acts on the characters of G. It leaves $\mathrm{Irr}(G)$ invariant, and the action is determined by its restriction to $\mathrm{Irr}(G)$. The factor group by the kernel of the action will be denoted by $\mathrm{Gal}(\mathrm{Irr}(G))$.

A character is called *real*, resp. *rational*, if all its values are real, resp. rational, numbers. A conjugacy class is called *real*, resp. *rational*, if its elements are conjugate to their inverses, resp. to all those n-th powers with n coprime to the element order. If every element in G is conjugate to its inverse then all characters and all conjugacy classes of G are real, and if every element in G is conjugate to all powers that are coprime to its order then all characters and all conjugacy classes of G are rational.

The sum of all distinct Galois conjugates of a character χ is clearly rational, and if χ is irreducible then this sum is called a *rational irreducible* character. We denote the set of rational irreducible characters of G by $\mathrm{Rat}(G)$

7. The Module $\mathcal{H}^1(X)$ of Holomorphic Differentials

In this section, we introduce a natural module for $\mathrm{Aut}(X)$. For that, we need the notion of holomorphic *differentials* (or *differential forms* or *1-forms*) on Riemann surfaces. Like the material in the previous chapter, this can be generalized. For the definition of not necessarily holomorphic differentials, see [**FK92**, I.3.1] or [**Mir95**, IV.1]. A geometric description of differentials in terms of cotangent spaces can be found, e.g., in [**For77**, §9].

DEFINITION 7.1. Let X be a Riemann surface, and $\mathcal{A}$ an atlas for the analytic structure on X. A *holomorphic differential* on X is a map ω that assigns to each chart (U, z) of $\mathcal{A}$ a holomorphic function $f_z \colon z(U) \to \mathbb{C}$, such that the following compatibility condition holds.

If (U_1, z_1) and (U_2, z_2) are two charts in $\mathcal{A}$ with $U_1 \cap U_2 \neq \emptyset$ then let $T = z_2 \circ z_1^{-1}$ be the coordinate change, which is a holomorphic function in z_1, defined on $z_1(U_1 \cap U_2)$. We require that

$$f_{z_1} = (f_{z_2} \circ T) \cdot \frac{dT}{dz_1}$$

on $z_1(U_1 \cap U_2)$.

When defining differentials like this, one has to show that it is possible to assign a compatible value for each chart that is compatible with $\mathcal{A}$ but does not belong to $\mathcal{A}$, see [**Mir95**, Lemma IV.1.4]. Conversely, by the identity theorem for holomorphic functions, a differential is defined already by its value on a single chart.

Usually one writes $\omega = f(z)\,dz$ for a holomorphic differential ω. For a compatible coordinate $\tilde{z} = T(z)$ with $\omega = \tilde{f}(\tilde{z})\,d\tilde{z}$, one gets formally that

$$f(z)\,dz = \left(\tilde{f}(T(z)) \cdot \frac{dT(z)}{dz} \right) dz = \tilde{f}(T(z))\,dT(z) = \tilde{f}(\tilde{z})\,d\tilde{z}.$$

From this viewpoint, a holomorphic differential is an assignment of a function f to each coordinate z such that the expression $f(z)\,dz$ is invariant under coordinate changes, see [**FK92**, I.3.1]

For example, assigning the zero function to each chart always defines a holomorphic differential. With the obvious definition of addition and scalar multiplication of differentials, one gets that the holomorphic differentials on a Riemann surface X form a complex vector space. This space is denoted by $\mathcal{H}^1(X)$.

THEOREM 7.2. *For a compact Riemann surface X, $\dim \mathcal{H}^1(X) = g(X)$.*

Proof. [**FK92**, III.5.2] $\square$

We want to define an action of $\mathrm{Aut}(X)$ on $\mathcal{H}^1(X)$. More generally, let $F\colon X \to Y$ be a biholomorphic map between two Riemann surfaces X and Y. Then F induces a vector space isomorphism $F^*\colon \mathcal{H}^1(Y) \to \mathcal{H}^1(X)$ via $F^*(\omega) = \omega \circ F$, that is, $\omega = f(z)\,dz \in \mathcal{H}^1(Y)$ is mapped to $F^*(\omega) = f(z \circ F)\,d(z \circ F) \in \mathcal{H}^1(X)$. To be precise, we choose an atlas $\mathcal{A}$ for the analytic structure on Y. For each chart (U, z) in $\mathcal{A}$, the chart $(F^{-1}(U), z \circ F)$ is compatible with the analytic structure on X, so $\{(F^{-1}(U), z \circ F) \mid (U, z) \in \mathcal{A}\}$ is an atlas for the analytic structure on X. Now for $\omega \in \mathcal{H}^1(Y)$, assigning $\omega(z)$ to the coordinate $z \circ F$ defines the holomorphic differential $F^*(\omega)$ on X.

We apply this to the case $Y = X$. For $\sigma \in \mathrm{Aut}(X)$, σ^* is the induced action on $\mathcal{H}^1(X)$. Because

$$(\sigma_1 \circ \sigma_2)^*(\omega) = \omega \circ (\sigma_1 \circ \sigma_2) = (\omega \circ \sigma_1) \circ \sigma_2 = \sigma_2^*(\sigma_1^*(\omega))$$

we have $(\sigma_1 \circ \sigma_2)^* = \sigma_2^* \circ \sigma_1^*$, so the map $\rho\colon \sigma \mapsto (\sigma^{-1})^*$ defines a linear action of $\mathrm{Aut}(X)$ on $\mathcal{H}^1(X)$.

REMARK 7.3. We have used special charts to define $\sigma^*(\omega)$ for $\sigma \in \mathrm{Aut}(X)$ and $\omega \in \mathcal{H}^1(X)$, namely (U, z) and $(\sigma^{-1}(U), z \circ \sigma)$. Note that σ is described by the identity map with respect to these charts. For arbitrary charts (U, z) and $(\sigma^{-1}(U), z')$, σ is described by $T = z \circ \sigma \circ (z')^{-1}$. Using the compatibility condition for differentials, we get the general formula

$$\sigma^*(\omega) = \omega(T(z')) \cdot \frac{dT}{dz'}\,dz',$$

as given in [**FK92**, V.2.1].

If $F\colon X \to Y$ is biholomorphic then the map $\hat{F}\colon \sigma \mapsto F^{-1} \circ \sigma \circ F$ is a group isomorphism from $\mathrm{Aut}(Y)$ to $\mathrm{Aut}(X)$, with the property that

$$F^* \circ \sigma^* = (\sigma \circ F)^* = (F \circ \hat{F}(\sigma))^* = \hat{F}(\sigma)^* \circ F^*.$$

Identifying $\mathrm{Aut}(Y)$ and $\mathrm{Aut}(X)$, we get that F^* is an isomorphism of modules for $\mathrm{Aut}(X)$. As a consequence, the character of the above representation is invariant under biholomorphic equivalence. This motivates the following definition, as introduced in [**KK90a, KK91**].

DEFINITION 7.4. Let χ be a character of the finite group G. We say that χ *comes from a Riemann surface* if there are a compact Riemann surface X of genus $g(X) \geq 2$ and an isomorphism $\varphi\colon \tilde{G} \to G$ for a group $\tilde{G} \leq \mathrm{Aut}(X)$ such that $\chi \circ \varphi$ is the character of the natural action of $\tilde{G}$ on $\mathcal{H}^1(X)$. We will identify the groups G and $\tilde{G}$.

A character that comes from a Riemann surface carries a lot of information about the underlying surface. One example is the following. (Note that in the proof, we use Corollary 12.3 which will be derived in Chapter 3.)

THEOREM 7.5. *Let X be a compact Riemann surface, and $G \leq \mathrm{Aut}(X)$. A differential $\omega \in \mathcal{H}^1(X)$ is called G-invariant if and only if $\sigma(\omega) = \omega$ for all $\sigma \in G$. The vector space of G-invariant differentials in $\mathcal{H}^1(X)$ has dimension $g(X/G)$.*

Proof. Following [**Sah69**], we give a character theoretic proof. For proofs using differentials themselves, see [**Lew63**, Theorem 2] or [**FK92**, V.2.2].

Let χ be the character afforded by the natural action of G on $\mathcal{H}^1(X)$. We have to show that $[\chi, 1_G] = g(X/G)$. By Theorem 7.2, $\chi(1) = g(X)$, and $\chi(\sigma) + \overline{\chi(\sigma)} = 2 - |\mathrm{Fix}_X(\sigma)|$ for all $\sigma \in G^\times$ by Corollary 12.3. Thus

$$|G| \cdot [\chi + \overline{\chi}, 1_G] = \sum_{\sigma \in G}(\chi(\sigma) + \overline{\chi(\sigma)}) = 2g + 2(|G| - 1) - \sum_{\sigma \in G^\times} |\mathrm{Fix}_X(\sigma)|.$$

Now

$$\sum_{\sigma \in G^\times} |\mathrm{Fix}_X(\sigma)| = |\{(x, \sigma) \in X \times G^\times \mid x^\sigma = x\}| = \sum_{x \in X}(|\mathrm{Stab}_G(x)| - 1)$$

yields

$$g - 1 = |G| \cdot \left(\frac{1}{2}[\chi + \overline{\chi}, 1_G] - 1\right) + \frac{1}{2}\sum_{x \in X}(|\mathrm{Stab}_G(x)| - 1)$$

by Lemma 4.4, and we get $g(X/G) = \frac{1}{2}[\chi + \overline{\chi}, 1_G] = [\chi, 1_G]$.   □

8. Characters that Come from Riemann Surfaces

It is already clear from Definition 7.4 that the property of coming from a Riemann surface is defined only up to group automorphisms. More precisely, if the G-character χ comes from a Riemann surface and $\sigma \in \mathrm{Aut}(G)$ then the character $\chi \circ \sigma$ also comes from a Riemann surface.

In Situation 3.22, the surface kernel epimorphism Φ identifies the automorphism group Γ/K with the abstract group G, so we may regard $\chi \circ \Phi$ as a character of Γ, with kernel K. As we have sketched in Section 5.2, the normal subgroups of Γ with factor group isomorphic to G correspond to $\mathrm{Aut}(G)$-orbits on $\mathrm{Epi}(\Gamma, G)$. In terms of characters this means that all characters in an $\mathrm{Aut}(G)$-orbit belong to the same normal subgroup of Γ and hence come from the same Riemann surface. (But note that different normal subgroups K of Γ may induce the same $\mathrm{Aut}(G)$-orbit of G-characters, see Example 15.31 below.)

The group $\mathrm{Gal}(\mathrm{Irr}(G))$ acts on the characters that come from Riemann surfaces. This is clear for those Galois automorphisms that can be realized by group automorphisms, for the other Galois automorphisms it will be shown in Section 15.7. As a consequence, $\mathrm{Gal}(\mathrm{Irr}(G))$ acts on the normal subgroups of Γ with factor group isomorphic to G, see Example 15.31.

If the group G is given only via its character table, for example by a table in the Atlas of Finite Groups [CCN+85], the irreducible characters of G are determined only up to *table automorphisms*, that is, those permutations of the columns of the table that leave the (irreducible) characters invariant and respect the power maps. Choosing one character from an orbit under table automorphisms means just a choice of an identification of the columns of the table with the conjugacy classes of the group.

But note that in general we cannot say that all characters in an orbit under table automorphisms come from Riemann surfaces if one character in the orbit does. That is, if χ comes from a Riemann surface then its orbit under table automorphisms may contain characters that do not come from Riemann surfaces, relative to the choice of χ. For more details, see Section 15.7 and Example 18.7.

CHAPTER 3

Automorphisms of Compact Riemann Surfaces

In Section 9, abelian surface kernel factors of Fuchsian groups are classified, mainly following [**Har66**]. This describes in particular all orders of automorphisms of Riemann surfaces of a fixed genus.

The following sections 10 and 11 deal with the relation between the cardinalities of certain sets of fixed points and surface kernel epimorphisms. Part of this material can be found in [**Kur84, KK91**].

With the notation defined in these sections, we can state the Eichler Trace Formula in Section 12, which will be, together with the Riemann–Hurwitz Formula, the most important tool in the subsequent chapters.

Finally, Section 13 gives a summary of the different levels of information about a surface kernel epimorphism that were developed before.

9. Abelian Groups of Automorphisms

A necessary and sufficient condition for determining from the signature of a Fuchsian group Γ the existence of a surface kernel homomorphism with image a prescribed *abelian* group is given by the following theorem. Both the statement and its proof are slight generalizations of [**Har66**, Theorem 4], which treats the case of cyclic groups. Some arguments in the proof use the facts about abelian invariants that are collected in Appendix A.

THEOREM 9.1. *Let Γ be as in Theorem 3.2, with $M = \mathrm{lcm}(m_1, m_2, \ldots, m_r)$. There is a surface kernel epimorphism from Γ onto a finite abelian group A if and only if the following conditions are satisfied.*

(o) *There exists an epimorphism from Γ onto A;*
(i) $\mathrm{lcm}(m_1, m_2, \ldots, m_{i-1}, m_{i+1}, \ldots, m_r) = M$ *for all i;*
(ii) M *divides the exponent $\exp(A)$ of A, and if $g = 0$, $M = \exp(A)$;*
(iii) $r \neq 1$, *and if $g = 0$, $r \geq 3$;*
(iv) *if M is even and only one of the abelian invariants of A is divisible by the maximum power of 2 dividing M, the number of m_i divisible by the maximum power of 2 dividing M is even.*

29

Proof. First we prove the necessity of the conditions. Let Φ be a surface kernel homomorphism that maps Γ onto an abelian group. Then $\prod_{i=1}^{r} \Phi(c_i) = 1$, so $\prod_{i \neq j} \Phi(c_i) = \Phi(c_j)^{-1}$ for each j. This implies $1 = \prod_{i \neq j} \Phi(c_i)^{\mathrm{lcm}_{i \neq j}(m_i)} = \Phi(c_j)^{-\mathrm{lcm}_{i \neq j}(m_i)}$, and since $\Phi(c_j)$ has order m_j, we conclude that m_j divides $\mathrm{lcm}_{i \neq j}(m_i)$. This is equivalent to condition (i).

The group generated by $\{\Phi(c_i) \mid 1 \leq i \leq r\}$ has exponent $\mathrm{lcm}(m_1, m_2, \ldots, m_r)$, which must divide $\exp(A)$. If $g = 0$, it must be the whole group A.

The condition $r \neq 1$ follows from $\prod_{i=1}^{r} \Phi(c_i) = 1$ and thus is a consequence of (i). If $g = 0$ and $r = 2$ then

$$(g - 1) + \frac{1}{2} \sum_{i=1}^{r} (1 - 1/m_i) = -\frac{1}{2}(1/m_1 + 1/m_2) < 0,$$

so Γ is not Fuchsian, by Theorem 3.12.

To deduce condition (iv) in the case of even M and a unique abelian invariant of A that is divisible by the 2-part of M, define the homomorphism Φ' from Γ onto an abelian group of exponent M by $\Phi'(c_i) = \Phi(c_i)$ and $\Phi'(a_i) = \Phi'(b_i) = 1$. Then consider the homomorphism Ψ from Γ onto a *cyclic* group of order 2, defined by $\Psi(x) = \Phi'(x)^{M/2}$. Since $\prod_{i=1}^{r} \Phi'(c_i) = 1$, the number of i with $\Psi(c_i)$ not the identity must be even. But $\Psi(c_i)$ is not the identity if and only if m_i is divisible by the largest power of 2 dividing M.

To prove the sufficiency of the conditions, we construct an appropriate surface kernel epimorphism. For that, suppose A is a direct product of p-groups A_p, for p in a set P of primes. It is sufficient to construct, for each $p \in P$, an epimorphism from Γ onto A_p whose kernel has no p-torsion. The direct product of these homomorphisms will then give an epimorphism with torsion-free kernel.

So we may assume that $M = p^\mu$ is a prime power, as well as that $m_i = p^{\mu_i}$, with $\mu_1 \leq \mu_2 \leq \cdots \leq \mu_r$. By condition (o), there is an epimorphism $\Phi_0 \colon \Gamma \to A$. It factors through the natural epimorphism onto Γ/Γ' whose image has by Lemma A.3 the abelian invariants $(0, 0, \ldots, 0, m_1, m_2, \ldots, m_{r-1})$, the number of zeros being $2g_0$. As used in the proof of Lemma A.3, the abelian invariants correspond to the hyperbolic generators of Γ and the first $r - 1$ elliptic generators.

Now we construct an epimorphism Φ with $|\Phi(c_i)| = m_i$ for all i, by modifying Φ_0. That is, we take a cyclic subgroup $\langle a \rangle$ of A of order m_r, and set $\Phi(c_i) = \Phi_0(c_i)a^{\xi_i}$, with exponents ξ_i such that a^{ξ_i} has order m_i. Note that this is possible because the largest abelian invariant of A is a multiple of all m_i by condition (ii). For $1 \leq i \leq r - 2$, we choose $\xi_i = p^{\mu - \mu_i}$, so a^{ξ_i} has order m_i, as required. Φ will be a homomorphism if and only if $\sum_{i=1}^{r} \xi_i \equiv 0 \pmod{m_r}$, so this is a condition on the choices of ξ_{r-1} and ξ_r.

Let t be the number of i with $\mu_i = \mu$, that is, $\mu_{r-t} < \mu_{r-t+1} = \mu_{r-t+2} = \cdots = \mu_r = \mu$. By condition (i), we have $t \geq 2$. Let $R = (1/p) \cdot \sum_{i=1}^{r-t} p^{\mu-\mu_i}$, which is an integer. Then

$$\sum_{i=1}^{r} \xi_i = \sum_{i=1}^{r-2} p^{\mu-\mu_i} + \xi_{r-1} + \xi_r = pR + t - 2 + \xi_{r-1} + \xi_r.$$

If $t \not\equiv 1 \pmod{p}$ then we set $\xi_{r-1} = 1$ and $\xi_r = 1 - t - pR$. If $t \equiv 1 \pmod{p}$ then we set $\xi_{r-1} = -1$ and $\xi_r = 3 - t - pR$. Then Φ is an epimorphism by construction, and it has torsion-free kernel except if $p = 2$ and t is odd.

But in the remaining situation, A has a subgroup with abelian invariants (m_r, m_r) by condition (iv), so we may multiply the images of c_{r-1} and c_r by mutually inverse elements of order m_r in a cyclic subgroup of A that intersects trivially with the subgroup spanned by the element a. $\qquad\square$

REMARK 9.2. Note that condition (o) is a condition on the abelian invariants of Γ/Γ', which can be computed from the signature of Γ by Lemma A.3. So we can check this condition using Lemmas A.1 and A.2.

EXAMPLE 9.3 (see [**Mac65**, Theorem 1]). A necessary condition on the existence of abelian surface kernel factors of $\Gamma = \Gamma(g; m_1, m_2, \ldots, m_r)$ is that the derived subgroup Γ' is torsion-free. This is equivalent to condition (i) of Theorem 9.1. To see this, observe that the conditions (o) and (iii) are trivially satisfied for $A = \Gamma/\Gamma'$, and that (ii) and (iv) follow from Lemma A.3.

Note that Γ/Γ' is infinite for $g > 0$. For proving the sufficiency of condition (i), we may apply Theorem 9.1 to get an epimorphism to the torsion part of the commutator factor group, and map the hyperbolic generators of Γ to independent generators of infinite order. $\qquad\diamond$

In the special case that the abelian group A is a cyclic group of order m, condition (o) is implied by the other conditions, so we get Harvey's Theorem as

COROLLARY 9.4. *Let Γ be as in Theorem 3.2, with $M = \operatorname{lcm}(m_1, m_2, \ldots, m_r)$. There is a surface kernel homomorphism from Γ onto a cyclic group of order m if and only if the following conditions are satisfied.*

(i) $\operatorname{lcm}(m_1, m_2, \ldots, m_{i-1}, m_{i+1}, \ldots, m_r) = M$ *for all i;*
(ii) *M divides m, and if $g = 0$, $M = m$;*
(iii) *$r \neq 1$, and if $g = 0$, $r \geq 3$;*
(iv) *if M is even, the number of m_i divisible by the maximum power of 2 dividing M is even.*

Using the existence theorem 3.12, we get the following classification of abelian automorphism groups. In particular, the orders of automorphisms of compact Riemann surfaces of given genus can be derived from it.

COROLLARY 9.5. *There is a compact Riemann surface of genus $g \geq 2$ with an abelian automorphism group isomorphic to A if and only if*

$$g - 1 = |A|(g_0 - 1) + \frac{|A|}{2} \sum_{i=1}^{r} (1 - 1/m_i)$$

with nonnegative integers g_0 and $m_1, m_2, \ldots, m_r$ that satisfy the conditions of Theorem 9.1.

As an application, we prove upper bounds for the orders of automorphisms and the orders of abelian groups of automorphisms. The former bound is derived in [**Wim95**] and [**Har66**, Corollary after Theorem 6], the latter follows by an easy computation from [**Mac65**, Theorem 4].

COROLLARY 9.6.   (a) *The maximum order of an automorphism of a compact Riemann surface of genus $g \geq 2$ is $4g + 2$.*

(b) *The maximum order for an abelian group of automorphisms of a compact Riemann surface of genus $g \geq 2$ is $4g + 4$.*

Proof. Suppose there is a compact Riemann surface of genus $g \geq 2$ that admits an abelian group A of automorphisms, and let $|A| = m$. Let g_0 denote the genus of the orbit space under A. According to the Riemann–Hurwitz Formula (Lemma 3.13), there are divisors $m_i \geq 2$ of m, $1 \leq i \leq r$, such that

$$g - 1 = m(g_0 - 1) + \frac{m}{2} \sum_{i=1}^{r} (1 - 1/m_i).$$

We have to show that $g - 1 \geq (m - 8)/4$ to prove part (b). For part (a), in the case that A is cyclic we must show the stronger inequality $g - 1 \geq (m - 6)/4$.

Since $g \geq 2$, we are done if $m \leq 10$, so we consider only groups of order $m > 10$. By part (a) of Lemma 3.18, we may also assume that $g_0 = 0$ and $3 \leq r \leq 4$.

If $r = 4$ and $m_2 \geq 3$,

$$g - 1 = m - \sum_{i=1}^{4} m/(2m_i) \geq m - (m/4 + 3m/6) = m/4 > (m - 6)/4.$$

Now A is generated by $r - 1$ out of the r generators of orders $m_1, m_2, \ldots, m_r$. So we have $m \leq m_1 m_2 \cdots m_r / m_i$, for any i.

For $r = 4$ and $m_1 = m_2 = 2$, this implies $m \leq 4m_3$ and $m \leq 4m_4$, thus

$$g - 1 = m/2 - m/(2m_3) - m/(2m_4) \geq m/2 - 4 > (m - 6)/4,$$

by our assumption that $m > 10$.

Finally, in the case $g_0 = 0$ and $r = 3$, we are done if $m_1 \geq 6$, since

$$g - 1 = m/2 - \sum_{i=1}^{3} m/(2m_i) \geq m/2 - 3m/12 = m/4 > (m-6)/4.$$

In the remaining cases, first we prove part (a) of the corollary. Here we use the condition that $m = \operatorname{lcm}(m_1, m_2) = \operatorname{lcm}(m_1, m_3) = \operatorname{lcm}(m_2, m_3)$, from which we get that m_2 and m_3 must be multiples of m/m_1, moreover at least m_3 is equal to m because m_1 is a prime power in all cases. So we have

$$\begin{aligned} g - 1 &= m/2 - m/(2m_1) - m/(2m_2) - m/(2m) \\ &\geq m/2 - m/(2m_1) - m_1/2 - 1/2. \end{aligned}$$

The right hand side is equal to $(m-6)/4$, $m/3-2$, $(3m-20)/8$, and $2m/5-3$, for $m_1 = 2, 3, 4, 5$, respectively, and all these are larger than or equal to $(m-6)/4$ for $m \geq 10$.

Now we prove part (b) for $g_0 = 0$, $r = 3$, and $m_1 \leq 5$. For that, we need to consider only $m > 12$. We use the inequalities $m \leq m_1 m_2$ and $m \leq m_1 m_3$, which yield

$$g - 1 = m/2 - \sum_{i=1}^{3} m/(2m_i) \geq m/2 - m/(2m_1) - m_1.$$

If $m_1 = 2$, this means $g - 1 \geq m/4 - 2 = (m-8)/4$. If $m_1 = 3$, it means $g - 1 \geq m/2 - m/6 - 3 = m/3 - 3$, which is larger than or equal to $(m-8)/4$ for $m \geq 12$. If $m_1 = 4$, we get $g - 1 \geq 3m/8 - 4$, which is larger than or equal to $(m-8)/4$ for $m \geq 16$; this is sufficient because 4 divides m. If $m_1 = 5$, we get $g - 1 \geq m/2 - m/10 - 5 = 2m/5 - 5$, which is larger than or equal to $(m-8)/4$ for $m \geq 20$; the exceptional case $m = 15$ would describe a cyclic group, and it has been shown above that this cannot occur for $g = 2$. $\square$

We note some applications of these results.

EXAMPLE 9.7. For a nonnegative integer g, consider the group $\Gamma(0; 2, 2g + 1, 2(2g + 1))$. The conditions of Corollary 9.4 guarantee the existence of an epimorphism onto a cyclic group of order $2(2g+1) = 4g+2$, with torsion-free kernel of orbit genus

$$1 + (4g+2)(0-1) + \frac{4g+2}{2}\left(1 - \frac{1}{2} + 1 - \frac{1}{2g+1} + 1 - \frac{1}{4g+2}\right)$$

$$= -4g - 1 + \frac{4g+2}{2}\left(\frac{5}{2} - \frac{3}{4g+2}\right) = -4g - 1 + \frac{10g+5}{2} - \frac{3}{2} = g.$$

This shows that the upper bound proved in Corollary 9.6 (a) is attained for each g. ◇

EXAMPLE 9.8 (see [**FK92**, V.1.11 and V.2.14]). One can show, in the same way as in the proof of Lemma 3.18, that automorphisms of *prime* order p satisfy $p \leq g$ or $p = g + 1$ or $p = 2g + 1$. The latter two possibilities are attained for the groups $\Gamma(0; p, p, p, p)$ and $\Gamma(0; p, p, p)$, respectively. Note that all periods must be equal to p.   ◇

EXAMPLE 9.9. For abelian groups also, the upper bound is attained for each g. Consider the groups $\Gamma = \Gamma(0; 2, 2(g+1), 2(g+1))$ and $G = \langle a \rangle \times \langle b \rangle$ where a and b are of orders 2 and $2(g + 1)$, respectively. The map $c_1 \mapsto a$, $c_2 \mapsto b$, $c_3 \mapsto ab^{-1}$ extends to a surface kernel epimorphism $\Gamma \to G$.

Let $m = 4(g + 1) = |G|$, then the kernel of this epimorphism has orbit genus

$$1 - m + (m/2)(1 - 1/2 + 1 - 2/m + 1 - 2/m) = 1 + m/4 - 2 = g.$$

If g is even then $\Gamma(0; 2, 2(g + 1), 2(g + 1))$ has no torsion-free *nonabelian* factor group of order $4g + 4$. That is, let the factor group G of order $4g + 4$ be generated by two elements x, y of order $2(g+1)$, such that xy is an involution. Then the intersection of $\langle x \rangle$ and $\langle y \rangle$ is $\langle x^2 \rangle = \langle y^2 \rangle$ of odd order $g+1$, which is centralized by xy. Since xy is the only involution in $\langle x^2, xy \rangle$, $\langle xy \rangle$ is normal and thus central in G. By the fact that xy is not contained in $\langle x \rangle$, we conclude that G is a direct product $\langle x \rangle \times \langle xy \rangle$.   ◇

EXAMPLE 9.10. Let Γ be a Fuchsian group with orbit genus g_0, and let the normal subgroup K be a Fuchsian surface group of orbit genus g. The quotient $G = \Gamma/K$ describes a group of automorphisms of the Riemann surface $\mathcal{U}/K$. Suppose that G acts fixed point freely on $\mathcal{U}/K$, that is, $\mathcal{U}/K$ is a smooth covering of $\mathcal{U}/\Gamma$. In this case Γ acts fixed point freely on $\mathcal{U}$. So Γ is also a surface group, it has signature $(g_0; -)$, and the Riemann–Hurwitz Formula yields $(g - 1)/|G| = g_0 - 1$.

If G is *cyclic* of order m then the existence conditions of Corollary 9.4 become empty. In other words, there is a Riemann surface of genus g with a fixed point free automorphism of order m if and only if m divides $g - 1$.   ◇

EXAMPLE 9.11. In Table 4, all signatures $(g_0; m_1, m_2, \ldots, m_r)$ of Fuchsian groups are listed that have a torsion-free normal subgroup of orbit genus $g = 2$ and finite abelian factor group A of order m.

The list can be obtained by simply computing for the possible indices $1 \leq m \leq 4g + 4 = 12$ and for $g_0 \in \{0, 1\}$ all those r-tuples $(m_1, m_2, \ldots, m_r)$ such that the m_i divide m and the Riemann–Hurwitz Formula holds, i.e.,

$$g - 1 = m(g_0 - 1) + \frac{m}{2} \sum_{i=1}^{r} (1 - 1/m_i).$$

Then all signatures are discarded that do not satisfy the conditions of Theorem 9.1.

m	$(g_0; m_1, m_2, \ldots, m_r)$	A
1	$(2; -)$	1
2	$(1; 2, 2)$	2
	$(0; 2, 2, 2, 2, 2, 2)$	2
3	$(0; 3, 3, 3, 3)$	3
4	$(0; 2, 2, 4, 4)$	4
	$(0; 2, 2, 2, 2, 2)$	2×2
5	$(0; 5, 5, 5)$	5
6	$(0; 3, 6, 6)$	2×3
	$(0; 2, 2, 3, 3)$	2×3
8	$(0; 2, 8, 8)$	8
10	$(0; 2, 5, 10)$	2×5
12	$(0; 2, 6, 6)$	$2 \times 2 \times 3$

TABLE 4. Signatures for Abelian Groups of Genus 2

As no automorphism of order 7 exists for compact Riemann surfaces of genus 2, we see that the bound of $84(g - 1)$ on the order of automorphism groups is not attained for $g = 2$, i.e., there is no Hurwitz group of order 84.

Furthermore, we see that the signature $(0; 3, 3, 9)$ cannot lead to a group of automorphisms in genus 2, because its order would be 9, hence the group would be abelian; in fact it would be cyclic because $m_3 = 9$. But the signature does not satisfy the conditions of Theorem 9.1. (Questions like this will be discussed in Section 17.)   ◇

10. Fixed Points of Automorphisms

In this section, we study how the fixed points of a group of automorphisms of a Riemann surface are described by a surface kernel epimorphism.

DEFINITION 10.1. Let X be a compact Riemann surface, $G \leq \mathrm{Aut}(X)$, $h \in G^{\times}$, and $H = \langle h \rangle$. We define

$\mathrm{Fix}_X(h) = \mathrm{Fix}_X(H) = \{x \in X \mid xh = x\}$, the set of fixed points of h (or H),
$\mathrm{Fix}_X^G(h) = \mathrm{Fix}_X^G(H) = \{x \in X \mid H = \mathrm{Stab}_G(x)\}$, the set of points whose stabilizer in G is exactly H,
$CY(G)$, the set of all nontrivial cyclic subgroups of G, and
$CY(G, H) = \{K \in CY(G) \mid H < K\}$, the set of those cyclic subgroups of G that contain H properly.

We are interested mainly in the cardinalities $|\mathrm{Fix}_X(H)|$ and $|\mathrm{Fix}_X^G(H)|$. As the following observation shows, the knowledge of $|\mathrm{Fix}_X(H)|$ for all $H \in CY(G)$ is equivalent to the knowledge of all $|\mathrm{Fix}_X^G(H)|$.

LEMMA 10.2. *If X is a compact Riemann surface of genus $g \geq 2$, and $H \leq G \leq \mathrm{Aut}(X)$, then*

$$|\mathrm{Fix}_X^G(H)| = |\mathrm{Fix}_X(H)| - \sum_{K \in CY(G,H)} |\mathrm{Fix}_X^G(K)|.$$

Proof. The stabilizers of points in X are cyclic by Lemma 4.2, so

$$\mathrm{Fix}_X^G(H) = \mathrm{Fix}_X(H) \Big\backslash \biguplus_{K \in CY(G,H)} \mathrm{Fix}_X^G(K).$$

$\square$

LEMMA 10.3. *Let X be a Riemann surface, $G \leq \mathrm{Aut}(X)$, and $\pi\colon X \to X/G$ the canonical projection. For $H \leq G$, we have*

$$|\mathrm{Fix}_X^G(H)| = [N_G(H){:}H] \cdot |\pi(\mathrm{Fix}_X^G(H))|.$$

Proof. Let $x \in \mathrm{Fix}_X^G(H)$. For $\sigma \in G$, the group $\mathrm{Stab}_G(x\sigma) = \mathrm{Stab}_G(x)^\sigma = \sigma^{-1}H\sigma$ of $x\sigma$ is equal to H if and only if $\sigma \in N_G(H)$. The fibre of $\pi(x)$ consists of $[G{:}H]$ points, of which exactly $[N_G(H){:}H]$ have H as full stabilizer. So $|\mathrm{Fix}_X^G(H)|$ is the product of this number with the number of fibres that contain points of $\mathrm{Fix}_X^G(H)$, which is $|\pi(\mathrm{Fix}_X^G(H))|$. $\square$

By the fact that $\mathrm{Fix}_X^G(H^\sigma) = (\mathrm{Fix}_X^G(H))\sigma$ for any $\sigma \in G$, the stabilizers of all points in one fibre under π are a conjugacy class of subgroups in G.

In Situation 3.22, the points in X with nontrivial stabilizer are exactly the points in the fibres of $z_i\Gamma$, $1 \leq i \leq r$. So we can interpret that statement of Lemma 10.3 in such a way that for a (cyclic) subgroup H of G, exactly $|\mathrm{Fix}_X^G(H)|/[N_G(H) : H] = |\pi(\mathrm{Fix}_X^G(H))|$ of these r fibres consist of the points whose stabilizer in G is a G-conjugate of H.

Now let us look more closely at the relation between surface kernel epimorphisms and sets of fixed points. The following lemma shows in particular that the values $|\mathrm{Fix}_X^G(H)|$ determine the signature of Γ. The converse is not true in general, since the distribution of the periods m_i to the conjugacy classes of cyclic subgroups of order m_i in G may not be unique. But, for example, if G is cyclic then the signature of Γ determines the values $|\mathrm{Fix}_X^G(H)|$ for all $H \leq G$.

LEMMA 10.4. *In Situation 3.22, let $H \leq G$ be cyclic of order m. Then*

$$|\text{Fix}_X(H)| = |N_G(H)| \cdot \sum_{\substack{1 \leq i \leq r \\ m \mid m_i \\ H \sim_G \langle \Phi(c_i)^{m_i/m} \rangle}} \frac{1}{m_i}$$

and

$$|\text{Fix}_X^G(H)| = [N_G(H){:}H] \cdot |\{i \mid 1 \leq i \leq r, H \sim_G \langle \Phi(c_i) \rangle\}|.$$

Proof. (See [**Kur87**, Remark 1.2].) By Corollary 4.3, $z_i K \in \text{Fix}_X^G(\Phi(c_i))$, so the number of i, $1 \leq i \leq r$, such that H is G-conjugate to $\langle \Phi(c_i) \rangle$ equals the number of fibres that consist of points whose full stabilizer is a G-conjugate of H. Thus Lemma 10.3 yields the second formula. To prove the first formula, let R_i be a set of coset representatives of $\text{Stab}_G(z_i K) = \langle \Phi(c_i) \rangle$ in G, for $1 \leq i \leq r$. Then

$$\text{Fix}_X(H) = \biguplus_{1 \leq i \leq r} \{(z_i K)^\sigma \mid \sigma \in R_i, H \leq \langle \Phi(c_i) \rangle^\sigma\}$$

$$= \biguplus_{1 \leq i \leq r} \{(z_i K)^\sigma \mid \sigma \in R_i, H = \langle \Phi(c_i)^{m_i/m} \rangle^\sigma\}.$$

Taking the cardinalities on both sides, we get

$$|\text{Fix}_X(H)| = \sum_{1 \leq i \leq r} |\{(z_i K)\sigma \mid \sigma \in R_i, H = \langle \Phi(c_i)^{m_i/m} \rangle^\sigma\}|$$

$$= \sum_{1 \leq i \leq r} |\{\sigma \in R_i \mid H = \langle \Phi(c_i)^{m_i/m} \rangle^\sigma\}|$$

$$= \sum_{1 \leq i \leq r} \frac{1}{m_i} |\{\sigma \in G \mid H = \langle \Phi(c_i)^{m_i/m} \rangle^\sigma\}|,$$

where the set in the i-th summand has cardinality $|N_G(H)|$ if H is G-conjugate to $\langle \Phi(c_i)^{m_i/m} \rangle$, and is empty otherwise. $\square$

EXAMPLE 10.5. Let $\Gamma = \Gamma(0; 2, 5, 10)$, and $G = \langle \sigma \rangle$ a cyclic group of order 10. Then $|\text{Fix}_X(\sigma)| = |\text{Fix}_X^G(\sigma)| = 1$, $|\text{Fix}_X^G(\sigma^2)| = 2$ and thus $|\text{Fix}_X(\sigma^2)| = 3$, $|\text{Fix}_X^G(\sigma^5)| = 5$, so $|\text{Fix}_X(\sigma^5)| = 6$.

The preimage of $\langle \sigma^2 \rangle$ in Γ has signature $(0; 5, 5, 5)$, that of $\langle \sigma^5 \rangle$ has signature $(0; 2, 2, 2, 2, 2, 2)$. ◇

EXAMPLE 10.6. Let p be an odd prime, $\Gamma = \Gamma(0; 2, 2, p, p)$, $G = \langle a, b \mid a^p, b^2, (ab)^2 \rangle$ the dihedral group of order $2p$, and define the epimorphism Φ by

$$\Phi(c_1) = \Phi(c_2) = b, \quad \Phi(c_3) = a, \quad \text{and} \quad \Phi(c_4) = a^{-1};$$

$\langle a \rangle$ is normal in G, $\langle b \rangle$ is self-normalizing, and both $\langle a \rangle$ and $\langle b \rangle$ are maximal subgroups of G. By Lemma 10.4,

$$|\text{Fix}_X(a)| = |\text{Fix}_X^G(a)| = [N_G(a){:}\langle a \rangle] \cdot |\{i \mid \langle a \rangle \sim_G \langle \Phi(c_i) \rangle\}| = 2 \cdot 2 = 4$$

and

$$|\text{Fix}_X(b)| = |\text{Fix}_X^G(b)| = [N_G(b){:}\langle b \rangle] \cdot |\{i \mid \langle b \rangle \sim_G \langle \Phi(c_i) \rangle\}| = 1 \cdot 2 = 2.$$

◇

11. Rotation Constants

In this section, we refine the results of the previous section by taking into account how an automorphism acts in a neighborhood of each of its fixed points. For that, we need to refine our notation.

DEFINITION 11.1. Let X be a compact Riemann surface, $h \in G \leq \text{Aut}(X)$ of order m, and $u \in \mathbb{Z}$ with $\gcd(u, m) = 1$. We define

$\text{Fix}_{X,u}(h) = \{x \in \text{Fix}_X(h) \mid \zeta_x(h) = \zeta_m^u\}$, the set of fixed points of h with rotation constant ζ_m^{-u},

$\text{Fix}_{X,u}^G(h) = \text{Fix}_X^G(h) \cap \text{Fix}_{X,u}(h)$, the set of points with stabilizer exactly $\langle h \rangle$ and with rotation constant of h equal to ζ_m^{-u}, and

$I(m) = \{u \mid 1 \leq u < m, \gcd(u, m) = 1\}$, the set of positive integers smaller than m that are coprime to m.

For these sets of fixed points, we get statements analogous to those of Section 10.

LEMMA 11.2. Let X be a Riemann surface, $h \in G \leq \text{Aut}(X)$ of order m, and $u \in I(m)$. For each group in $CY(G, \langle h \rangle)$, choose a generator k such that $k^d = h$ for $d = |k|/|h|$. Denote the set of these k by $cy(G, h)$. Then

$$|\text{Fix}_{X,u}^G(h)| = |\text{Fix}_{X,u}(h)| - \sum_{k \in cy(G,h)} \sum_{\substack{v \in I(|k|) \\ v \equiv u \ (\text{mod } m)}} |\text{Fix}_{X,v}^G(k)|.$$

Proof. By the proof of Lemma 10.2, we have

$$\text{Fix}_X^G(h) = \text{Fix}_X(h) \Big\backslash \biguplus_{k \in cy(G,h)} \text{Fix}_X^G(k)$$

and thus

$$\text{Fix}_{X,u}^G(h) = \text{Fix}_{X,u}(h) \Big\backslash \biguplus_{k \in cy(G,h)} \left(\text{Fix}_X^G(k) \cap \text{Fix}_{X,u}(h) \right).$$

Suppose k has order md, then $\text{Fix}_X^G(k) \cap \text{Fix}_{X,u}(h) =$

$$\{x \in \text{Fix}_X^G(k) \mid \zeta_x(h) = \zeta_m^u\}$$
$$= \{x \in \text{Fix}_X^G(k) \mid \zeta_x(k) = \zeta_{md}^v \quad \text{for} \quad v \in I(md) \quad \text{with} \quad (\zeta_{md}^v)^d = \zeta_m^u\}$$
$$= \biguplus_{\substack{v \in I(md) \\ v \equiv u \ (\text{mod } m)}} \text{Fix}_{X,v}^G(k).$$

$\square$

LEMMA 11.3. *Let X be a Riemann surface, $G \le \text{Aut}(X)$, $\pi \colon X \to X/G$ the canonical projection, $h \in G$ of order m and $u \in \mathbb{Z}$ with $\gcd(u, m) = 1$.*

We have $\text{Fix}_{X,u}^G(h^\sigma) = (\text{Fix}_{X,u}^G(h))\sigma$, in particular $|\text{Fix}_{X,u}(h)|$ depends only on the G-conjugacy class of h.

Furthermore, $\text{Fix}_{X,u}^G(h) = \text{Fix}_{X,uk}^G(h^k)$ for each $k \in I(m)$, and thus $h \sim_G h^k$ implies $|\text{Fix}_{X,u}^G(h)| = |\text{Fix}_{X,uk}^G(h)|$.

Proof. Let $\sigma \in G$, and φ be a local coordinate in $x_0 \in \text{Fix}_{X,u}^G(h)$. Then $\varphi \circ \sigma^{-1}$ is a local coordinate in $x_0\sigma$, with $\zeta_{x_0\sigma}(h^\sigma) = \zeta_{x_0}(h) = \zeta_m^u$. This means that $\text{Fix}_{X,u}^G(h^\sigma) = (\text{Fix}_{X,u}^G(h))\sigma$.

For the second statement, $\zeta_{x_0}(h^k) = \zeta_m^{uk}$ yields $\text{Fix}_{X,u}^G(h) \subseteq \text{Fix}_{X,uk}^G(h^k)$. The other inclusion follows from the fact that $k \in I(m)$ is invertible modulo m. $\square$

LEMMA 11.4. *Let X be a Riemann surface, $G \le \text{Aut}(X)$, and $\pi \colon X \to X/G$ the canonical projection. For $h \in G$ of order m and $u \in \mathbb{Z}$ with $\gcd(u, m) = 1$, we have*

$$|\text{Fix}_{X,u}^G(h)| = [C_G(h) \colon \langle h \rangle] \cdot |\pi(\text{Fix}_{X,u}^G(h))|.$$

Proof. If $x_0 \in \text{Fix}_{X,u}^G(h)$ then $[N_G(\langle h \rangle) \colon \langle h \rangle]$ points in the same fibre as x_0 lie in $\text{Fix}_X^G(h)$. By the first part of Lemma 11.3, $[C_G(h) \colon \langle h \rangle]$ of these points lie in $\text{Fix}_{X,u}^G(h)$, and by the second part of this lemma, h has different rotation constants at the other points. $\square$

LEMMA 11.5 (cf. [**Har71**, Theorem 7]). *In Situation 3.22, let $h \in G^\times$ be of order m and $u \in I(m)$. Then*

$$|\text{Fix}_{X,u}(h)| = |C_G(h)| \cdot \sum_{\substack{1 \le i \le r \\ m \mid m_i \\ h \sim_G \Phi(c_i)^{m_i u/m}}} \frac{1}{m_i}$$

and

$$|\text{Fix}_{X,u}^G(h)| = [C_G(h) \colon \langle h \rangle] \cdot |\{i \mid 1 \le i \le r, h \sim_G \Phi(c_i)^u\}|.$$

Proof. Set $P = (z_i K)\sigma$ for $\sigma \in G$. By the choice of the elliptic generators c_i, we have $\zeta_P(\Phi(c_i)^\sigma) = \zeta_{m_i}$, so $\zeta_P((\Phi(c_i)^{m_i u/m})^\sigma) = \zeta_{m_i}^{m_i u/m} = \zeta_m^u$. Similarly to the proof of Lemma 10.4, we derive the first equality from

$$\text{Fix}_{X,u}(h) = \biguplus_{1 \le i \le r} \{(z_i K)\sigma \mid \sigma \in R_i, h = (\Phi(c_i)^{m_i u/m})^\sigma\},$$

and the second from Lemma 11.4. $\square$

EXAMPLE 11.6. Let $G = L_3(2)$. We have $|G| = 168$, and G is generated by the matrices

$$\begin{pmatrix} 1 & 1 & 0 \\ 0 & 1 & 0 \\ 0 & 0 & 1 \end{pmatrix}, \begin{pmatrix} 0 & 0 & 1 \\ 1 & 0 & 0 \\ 0 & 1 & 0 \end{pmatrix}.$$

Their orders are 2 and 3, respectively, and their product has order 7. So there is a surface kernel epimorphism $\Phi \colon \Gamma(0; 2, 3, 7) \to G$, and $X = \mathcal{U}/\ker(\Phi)$ is a compact Riemann surface of genus

$$g(X) = 1 - |G| + \frac{|G|}{2}(1 - 1/2 + 1 - 1/3 + 1 - 1/7) = 3$$

with G as full group of automorphisms.

As in [CCN+85], we denote the conjugacy classes of G by 1A, 2A, 3A, 4A, 7A, 7B, and we choose the class names such that the product of the above generators lies in class 7A. By Lemma 11.5, the rotation constants of nonidentity elements in G are given by

$$
\begin{aligned}
|\text{Fix}_{X,1}(2A)| &= \tfrac{1}{2}|C_G(2A)| & & & & & &= 4, \\
|\text{Fix}_{X,1}(3A)| &= |\text{Fix}_{X,2}(3A)| &= \tfrac{1}{3}|C_G(3A)| & & & & &= 1, \\
|\text{Fix}_{X,1}(4A)| &= |\text{Fix}_{X,3}(4A)| & & & & & &= 0, \\
|\text{Fix}_{X,1}(7A)| &= |\text{Fix}_{X,2}(7A)| &= |\text{Fix}_{X,4}(7A)| &= \tfrac{1}{7}|C_G(7A)| &= 1, \\
|\text{Fix}_{X,3}(7A)| &= |\text{Fix}_{X,5}(7A)| &= |\text{Fix}_{X,6}(7A)| & & & & &= 0, \\
|\text{Fix}_{X,1}(7B)| &= |\text{Fix}_{X,2}(7B)| &= |\text{Fix}_{X,4}(7B)| & & & & &= 0, \\
|\text{Fix}_{X,3}(7B)| &= |\text{Fix}_{X,5}(7B)| &= |\text{Fix}_{X,6}(7B)| &= \tfrac{1}{7}|C_G(7B)| &= 1.
\end{aligned}
$$

The fixed points lie in three fibres. One contains $84 = 168/2$ points, 4 for each of the $21 = 168/8$ involutions in G. The second contains $56 = 168/3$ points, 2 for each of the $28 = 168/6$ subgroups of order 3 in G. Since the class 3A is rational, each element of order 3 has one fixed point with rotation constant ζ_3 and one with rotation constant ζ_3^2. The third fibre contains $24 = 168/7$ points, 3 for each of the $8 = 168/21$ subgroups of order 7 in G. Since an element σ of order 7 is conjugate to σ^2 and σ^4 in G, the numbers of fixed points with rotation constants ζ_7, ζ_7^2, and ζ_7^4 for σ are equal, and the numbers of fixed points with rotation constants ζ_7^3, ζ_7^5, and ζ_7^6 are equal. $\diamond$

12. The Eichler Trace Formula

A very important property of characters that come from Riemann surfaces is the following.

THEOREM 12.1 (Eichler Trace Formula). *Let σ be an automorphism of order $m > 1$ of a compact Riemann surface X of genus $g \geq 2$, and χ the character of the action of $\mathrm{Aut}(X)$ on $\mathcal{H}^1(X)$. Then*

$$\chi(\sigma) = 1 + \sum_{u \in I(m)} |\mathrm{Fix}_{X,u}(\sigma)| \frac{\zeta_m^u}{1 - \zeta_m^u}.$$

Proof. [**FK92**, V.2.4–V.2.9] □

In later chapters, the following property of the values $\frac{\zeta_m^u}{1-\zeta_m^u}$ will be used frequently.

LEMMA 12.2 (see [**FK92**, V.2.9]). *Let $m > 1$, $u \in I(m)$, and $\zeta = \zeta_m^u$. Then*

$$\frac{\zeta}{1-\zeta} + \frac{\zeta^{-1}}{1-\zeta^{-1}} = -1.$$

Proof. Compute

$$\frac{\zeta}{1-\zeta} + \frac{\zeta^{-1}}{1-\zeta^{-1}} = \frac{\zeta(1-\zeta^{-1}) + \zeta^{-1}(1-\zeta)}{1-\zeta-\zeta^{-1}+\zeta\zeta^{-1}} = \frac{\zeta - 1 + \zeta^{-1} - 1}{1 - \zeta - \zeta^{-1} + 1} = -1.$$

□

COROLLARY 12.3 (Lefschetz Fixed Point Formula). *With X, σ, and χ as in Theorem 12.1, we have*

$$\chi(\sigma) + \overline{\chi(\sigma)} = 2 - |\mathrm{Fix}_X(\sigma)|.$$

Proof. $\chi(\sigma) + \overline{\chi(\sigma)} =$

$$2 + \sum_{u \in I(m)} |\mathrm{Fix}_{X,u}(\sigma)| \left(\frac{\zeta_m^u}{1-\zeta_m^u} + \frac{\zeta_m^{-u}}{1-\zeta_m^{-u}} \right) = 2 - \sum_{u \in I(m)} |\mathrm{Fix}_{X,u}(\sigma)|.$$

□

As a consequence, we see that a character that comes from a Riemann surface is faithful, i.e., the action of $\mathrm{Aut}(X)$ on $\mathcal{H}^1(X)$ is faithful. Note that if χ is known to be real then its values are determined by the numbers $|\mathrm{Fix}_X(\sigma)|$, for $\sigma \in \mathrm{Aut}(X)^\times$.

REMARK 12.4. By Corollary 12.3, the character $\chi + \overline{\chi}$ is integral. In fact a stronger condition holds (see [**Sah69**, p. 34] and [**FK92**, III.2.7, V.3.1]), namely, the representation on the module $\mathcal{H}^1(X) \oplus \overline{\mathcal{H}^1(X)}$ is integral with respect to a suitable basis, where $\overline{\mathcal{H}^1(X)}$ is the space of antiholomorphic abelian differentials on X.

EXAMPLE 12.5. We classify the characters of fixed point free actions.

Let X be a compact Riemann surface X of genus at least 2, and χ the character of the action of $\mathrm{Aut}(X)$ on $\mathcal{H}^1(X)$. If $\sigma \in \mathrm{Aut}(X)^\times$ acts fixed point freely then $|\mathrm{Fix}_{X,u}(\sigma)| = 0$ for all u, so $\chi(\sigma) = 1$ by Theorem 12.1. Conversely, if $\chi(\sigma) = 1$ then Corollary 12.3 implies that $|\mathrm{Fix}_X(\sigma)| = 0$.

This means that a nonidentity element $\sigma \in \mathrm{Aut}(X)$ acts fixed point freely on X if and only if $\chi(\sigma) = 1$.

Suppose G is a fixed point free subgroup of $\mathrm{Aut}(X)$. Then the projection $X \to X/G$ is a smooth covering, the Riemann–Hurwitz Formula (Lemma 3.13) yields $\chi(1) = g(X) = 1 + |G|\,(g(X/G) - 1)$. Thus χ is the sum of the trivial character 1_G and $(g(X/G) - 1)$ times the regular character ρ_G of G. (This result was shown already in [**CW34**, p. 361], of course without using the Eichler Trace Formula.)

In terms of Fuchsian groups, the covering $X \to X/G$ is described by a factor $\Gamma/K \cong G$ of two surface groups $\Gamma \cong \Gamma(g(X/G); -)$ and $K \cong \Gamma(g(X); -)$.

So a character of the form $1_G + r \cdot \rho_G$ comes from a Riemann surface if G can be generated by $r + 1$ elements, as the proof of Corollary 3.15 shows. Analogously, if G *cannot* be generated by $2r + 2$ elements then $1_G + r \cdot \rho_G$ does *not* come from a Riemann surface, since then G cannot be an epimorphic image of $\Gamma(r+1; -)$. If G can be generated by $2r+2$ but not by $r+1$ elements then it depends on G whether $1_G + r \cdot \rho_G$ comes from a Riemann surface. ◇

We see that successively adding regular characters to a character $1_G + r \cdot \rho_G$ that comes from a Riemann surface again yields characters that come from Riemann surfaces. This observation can be generalized as follows.

EXAMPLE 12.6. Let $\Gamma = \Gamma(g_0; m_1, m_2, \dots, m_r)$, and $\Phi \colon \Gamma \to G$ be a surface kernel epimorphism, with χ the character of the induced action of G on $\mathcal{H}^1(\mathcal{U}/\ker(\Phi))$.

Γ is a factor group of $\Gamma' = \Gamma(g_0 + 1; m_1, m_2, \dots, m_r)$. Define $\Phi' \colon \Gamma' \to G$ as the composition of Φ and the natural epimorphism $\Gamma' \to \Gamma$ that maps the two new hyperbolic generators to the identity of Γ. Then Φ' is also a surface kernel epimorphism. Let χ' denote the character of the induced action of G on $\mathcal{H}^1(\mathcal{U}/\ker(\Phi'))$.

By Theorem 3.21, the generators of Γ and Γ' are chosen such that G acts on the two Riemann surfaces with the same rotation constants in the fixed points, so χ and χ' are equal on all nonidentity elements of G. The degree of χ' is $g(\mathcal{U}/\ker(\Phi')) = g(\mathcal{U}/\ker(\Phi)) + |G|$, hence $\chi' = \chi + \rho_G$. $\diamond$

EXAMPLE 12.7. Using Theorem 12.1, we compute the character χ associated to the action of the group $L_3(2)$ on $\mathcal{H}^1(X)$ for the surface X introduced in Example 11.6.

$$
\begin{aligned}
\chi(1A) &= g(X) && && = && 3, \\
\chi(2A) &= 1 + 4\frac{-1}{1-(-1)} && && = && -1, \\
\chi(3A) &= 1 + \frac{\zeta_3}{1-\zeta_3} + \frac{\zeta_3^2}{1-\zeta_3^2} && && = && 0, \\
\chi(4A) &= && && = && 1, \\
\chi(7A) &= 1 + \frac{\zeta_7}{1-\zeta_7} + \frac{\zeta_7^2}{1-\zeta_7^2} + \frac{\zeta_7^4}{1-\zeta_7^4} &&= \zeta_7 + \zeta_7^2 + \zeta_7^4 &&= b_7, \\
\chi(7B) &= 1 + \frac{\zeta_7^3}{1-\zeta_7^3} + \frac{\zeta_7^5}{1-\zeta_7^5} + \frac{\zeta_7^6}{1-\zeta_7^6} &&= \zeta_7^3 + \zeta_7^5 + \zeta_7^6 &&= b_7^*,
\end{aligned}
$$

where b_7 and b_7^* are the names used for the above values throughout [**CCN+85**].

So $\chi = \chi_2$ in the character table of $L_3(2)$ given in [**CCN+85**, p. 3].

Note that if we had chosen the image of the elliptic generator of order 7 in the class 7B instead of 7A then we would have got $\chi = \chi_3$. $\diamond$

13. Summary

We have introduced several levels of information to describe an automorphism group G of a compact Riemann surface X of genus g, namely

- a surface kernel epimorphism $\Phi \colon \Gamma \to G$,
- the numbers $|\mathrm{Fix}^G_{X,u}(h)|$ for $h \in G^\times$ and $u \in I(|h|)$,
- the numbers $|\mathrm{Fix}_{X,u}(h)|$ for $h \in G^\times$ and $u \in I(|h|)$,
- the character χ of the action of G on $\mathcal{H}^1(X)$,
- the numbers $|\mathrm{Fix}^G_X(H)|$ for $H \in CY(G)$,
- the numbers $|\mathrm{Fix}_X(H)|$ for $H \in CY(G)$,
- and the signature of the Fuchsian group Γ.

Each level contains the information of the levels below. More precisely,

- a surface kernel epimorphism determines the $|\mathrm{Fix}^G_{X,u}(h)|$ by Lemma 11.5,
- the knowledge of the $|\mathrm{Fix}^G_{X,u}(h)|$ is equivalent to the knowledge of the $|\mathrm{Fix}_{X,u}(h)|$ by Lemma 11.2,
- the $|\mathrm{Fix}_{X,u}(h)|$ determine the character χ by Theorem 12.1,
- from χ one can compute the values $|\mathrm{Fix}_X(H)|$ by Corollary 12.3,

- the knowledge of the $|\text{Fix}_X^G(h)|$ is equivalent to the knowledge of the $|\text{Fix}_X(h)|$ by Lemma 10.2,
- and the $|\text{Fix}_X^G(H)|$ determine the signature of Γ by Lemma 10.4.

The signature of Γ describes the orders of the images of the elliptic generators c_i under possible surface kernel epimorphisms $\Phi \colon \Gamma \to G$, the $|\text{Fix}_X(H)|$ describe the distribution of the c_i to the conjugacy classes of cyclic subgroups of G, and the $|\text{Fix}_{X,u}(h)|$ describe the distribution of the c_i to the conjugacy classes of elements in G.

In special cases, one can say more. For example, if the character table of G is real then the $|\text{Fix}_X(H)|$ determine the character χ, if G is abelian then the $|\text{Fix}_{X,u}^G(h)|$ determine a unique surface kernel epimorphism, and if G is cyclic then the signature of Γ determines the $|\text{Fix}_X^G(H)|$.

As a result of all this, the following definition makes sense.

DEFINITION 13.1. Let χ be the character of the action of G as subgroup of $\text{Aut}(X)$ on $\mathcal{H}^1(X)$ described by a surface kernel epimorphism Φ. We call χ the character *induced by* Φ, and denote it by $\chi = \text{Tr}(\Phi)$.

If a character χ defines the signature of a Fuchsian group Γ then we write $\Gamma = \Gamma(\chi)$.

CHAPTER 4

Generation of Groups

This chapter deals with the question whether a given finite group can be generated by elements with certain properties. Section 14 states a formula for the number of homomorphisms from a group with signature to a finite group, Section 15 collects criteria of generation and nongeneration, Section 16 shows an application concerning automorphism groups of Riemann surfaces, and Section 17 reinterprets the criteria in terms of signatures.

14. Structure Constants

Given a finite group G and a group

$$\Gamma = \left\langle a_1, b_1, a_2, b_2, \ldots, a_g, b_g, c_1, c_2, \ldots, c_r \;\middle|\; c_1^{m_1}, c_2^{m_2}, \ldots, c_r^{m_r}, \prod_{i=1}^{g}[a_i, b_i]\prod_{j=1}^{r} c_j \right\rangle$$

as in Theorem 3.2, we are interested in surface kernel epimorphisms $\Phi \colon \Gamma \to G$. In this section, we present Theorem 3 of [**Jon95**], which is a character theoretic condition on the existence and the number of certain homomorphisms $\Gamma \to G$. Later we will turn to the question of their surjectivity.

Let $\mathrm{Hom}(\Gamma, G)$ denote the set of homomorphisms from Γ to G, and $\mathrm{Epi}(\Gamma, G)$ the subset of surjective homomorphisms, as used already in Section 5.2. Identifying $\Phi \in \mathrm{Hom}(\Gamma, G)$ with the vector

$$(\Phi(a_1), \Phi(b_1), \Phi(a_2), \Phi(b_2), \ldots, \Phi(a_g), \Phi(b_g), \Phi(c_1), \Phi(c_2), \ldots, \Phi(c_r))$$

of images of the generators of Γ, we can regard $\mathrm{Hom}(\Gamma, G)$ as the set of solutions $\alpha_1, \beta_1, \alpha_2, \beta_2, \ldots, \alpha_g, \beta_g, \gamma_1, \gamma_2, \ldots, \gamma_r$ in G of the system of equations

$$\gamma_1^{m_1} = \gamma_2^{m_2} = \cdots = \gamma_r^{m_r} = \prod_{i=1}^{g}[\alpha_i, \beta_i]\prod_{j=1}^{r}\gamma_j = 1.$$

One property of surface kernel homomorphisms is that they respect the orders of the elliptic generators c_1, c_2, $\ldots$, c_r. To impose a more specific condition, we may prescribe a union L_i of conjugacy classes of elements of order m_i in G as set of admissible images of c_i. For $L = (L_1, \ldots, L_r)$, we denote by $\mathrm{Hom}_L(\Gamma, G)$ the set of homomorphisms that respect L in the sense that $\Phi(c_i) \in L_i$ for $\Phi \in \mathrm{Hom}_L(\Gamma, G)$, and we set $\mathrm{Epi}_L(\Gamma, G) = \mathrm{Epi}(\Gamma, G) \cap$

45

$\text{Hom}_L(\Gamma, G)$. We will also use the notations $\text{Hom}_L(g, G)$ and $\text{Epi}_L(g, G)$. If each L_i is a single conjugacy class then L is called a *class structure* of G.

The cardinality $|\text{Hom}_L(g, G)|$ can be computed from the character table of G by

THEOREM 14.1 (see [**Jon95**, Theorem 3]). *With L as above,*

$$|\text{Hom}_L(g, G)| = |G|^{2g-1} \sum_{\chi \in \text{Irr}(G)} \chi(1)^{2-2g-r} \prod_{i=1}^{r} \sum_{\sigma_i \in L_i} \chi(\sigma_i).$$

Proof. It is sufficient to prove this for class structures L. Taking disjoint unions will then yield the general result.

For $g = 0$ the equation reduces to

$$(*) \qquad |\text{Hom}_L(0, G)| = |G|^{-1} \sum_{\chi \in \text{Irr}(G)} \chi(1)^{2-r} \prod_{i=1}^{r} |L_i| \chi(\sigma_i)$$

where $\sigma_i \in L_i$. This is true by [**Mat87**, Satz II.6.1]. The special case (and induction start) $r = 3$ is treated also, e.g., in [**Isa76**, p. 45].

Now assume $g > 0$. We first choose class structures $K = (K_1, \dots, K_g)$ and $M = (K_1^{-1}, K_1, K_2^{-1}, K_2, \dots, K_g^{-1}, K_g, L_1, L_2, \dots, L_r)$ of G, where K_i consists of elements of order k_i, and K_i^{-1} of the inverses of K_i. By $[\alpha, \beta] = \alpha^{-1}\alpha^\beta = \alpha^{-1}(\beta^{-1}\alpha\beta)$, and with the notation

$$\text{Hom}_{K,L}(g, G) = \{\Phi \in \text{Hom}_L(g, G) \mid \Phi(a_i) \in K_i\},$$

the map

$$\text{Hom}_{K,L}(g, G) \qquad \to \qquad \text{Hom}_M(0, G)$$
$$(\alpha_1, \beta_1, \dots, \alpha_g, \beta_g, \gamma_1, \dots, \gamma_r) \mapsto (\alpha_1^{-1}, \alpha_1^{\beta_1}, \dots, \alpha_g^{-1}, \alpha_g^{\beta_g}, \gamma_1, \dots, \gamma_r)$$

is well-defined and surjective. Of course it is not injective; to be precise, we get the element $\beta_i^{-1}\alpha_i\beta_i$ with multiplicity $|C_G(\alpha_i)| = |G|/|K_i|$ as β_i ranges over G, thus

$$|\text{Hom}_{K,L}(g, G)| = \prod_{i=1}^{g} \frac{|G|}{|K_i|} |\text{Hom}_M(0, G)|.$$

By $(*)$, $|\text{Hom}_M(0, G)| =$

$$|G|^{-1} \sum_{\chi \in \text{Irr}(G)} \chi(1)^{2-(2g+r)} \prod_{j=1}^{g} |K_j|^2 \chi(\kappa_j)\chi(\kappa_j^{-1}) \prod_{i=1}^{r} |L_i|\chi(\sigma_i),$$

with $\sigma_i \in L_i$ and $\kappa_j \in K_j$. Hence $|\text{Hom}_{K,L}(g, G)| =$

$$|G|^{g-1} \sum_{\chi \in \text{Irr}(G)} \chi(1)^{2-(2g+r)} \prod_{i=1}^{r} |L_i|\chi(\sigma_i) \prod_{j=1}^{g} \sum_{\kappa_j \in K_j} \chi(\kappa_j)\chi(\kappa_j^{-1}).$$

Now $\mathrm{Hom}_L(g, G)$ is the disjoint union of the $\mathrm{Hom}_{K,L}(g, G)$, where K ranges over all g-tuples of G-classes. This summation over the last factors

$$\prod_{j=1}^{g} \sum_{\kappa_j \in K_j} \chi(\kappa_j)\chi(\kappa_j^{-1})$$

is equal to

$$\prod_{j=1}^{g} \sum_{\sigma_j \in G} \chi(\sigma_j)\chi(\sigma_j^{-1}) = \prod_{j=1}^{g} |G|[\chi, \chi] = |G|^g,$$

so we are done. $\qquad\square$

REMARK 14.2. The formula shows that the cardinality of $\mathrm{Hom}_L(g, G)$ is an invariant of the *unordered* vector L, which is clear by Lemma 3.3. By the argument of this lemma, $|\mathrm{Epi}_L(g, G)|$ also does not depend on the ordering of L.

Whenever *structure constants* occur in the following, we will use $|\mathrm{Hom}_L(g, G)|$. Note that different notions of structure constants are usual in the literature. The *ordinary structure constant* of a class structure $C = (C_1, C_2, \ldots, C_r)$ of G is defined as

$$\left|\left\{(\sigma_1, \sigma_2, \ldots, \sigma_{r-1}) \in C_1 \times C_2 \times \cdots \times C_{r-1} \mid \sigma_1\sigma_2\cdots\sigma_{r-1} = \sigma_r^{-1}\right\}\right|$$

for a fixed element $\sigma_r \in C_r$, so it is equal to $|\mathrm{Hom}_C(0, G)|/[G{:}C_G(\sigma_r)]$.

In [Mat87, p. 114], the *normalized structure constant* of C is defined as $|\mathrm{Hom}_C(0, G)|/[G{:}Z(G)]$.

15. Criteria of Generation and Nongeneration

This section collects some criteria for proving or disproving that $\mathrm{Epi}_L(g, G)$ is empty. The most interesting case will be $g = 0$.

We say that the group G is $(m_1, m_2, \ldots, m_r)$-*generated* if there is a class structure $C = (C_1, C_2, \ldots, C_r)$, with C_i consisting of elements of order m_i, such that $\mathrm{Epi}_C(0, G) \neq \emptyset$. In this case, G is also called an $(m_1, m_2, \ldots, m_r)$-*group*.

In [CWW92, Mat87], the notion of C-generation is introduced. According to the definition used there, G is C-generated if $\mathrm{Epi}_{C'}(0, G) \neq \emptyset$ for $C' = (C_1, C_2, \ldots, C_{r-1}, C_r')$, where C_r' consists of the inverses of the elements in C_r. But this should not cause confusion in the following.

15.1. Elementary Observations. We start with an immediate consequence of Theorem 14.1.

LEMMA 15.1 (see [**CWW92**, Lemma 1]). *Let G and L be as in Theorem 14.1, U an overgroup of G, $K = (K_1, K_2, \ldots, K_r)$ where K_i is a union of U-classes such that $L_i \subseteq K_i$. Furthermore, let S be a group that acts on U via conjugation and leaves each of the sets K_i invariant. If $|\mathrm{Hom}_K(g, U)| < [S{:}C_S(G)]$ then $\mathrm{Epi}_L(g, G) = \emptyset$.*

Proof. Assume that $\mathrm{Epi}_L(g, G) \neq \emptyset$, and choose $\varphi \in \mathrm{Epi}_L(g, G)$. For $s \in S$, let $\tau_s \in \mathrm{Aut}(S)$ denote the conjugation with s, and consider the orbit $M = \{\tau_s \circ \varphi \mid s \in S\}$. Clearly $s \in S$ acts on M via composition with τ_s, and $M \subseteq \mathrm{Hom}_K(g, U)$ because $L_i^{\tau_s} \subseteq K_i$ and $G^{\tau_s} \subseteq U^{\tau_s} \subseteq U$. The stabilizer of φ in S is $C_S(G)$, since $C_S(G)$ fixes φ and $\tau_s \circ \varphi = \varphi$ implies that s fixes a generating system of G pointwise. So we get $[S{:}C_S(G)] = |M| \leq |\mathrm{Hom}_K(g, U)|$. $\square$

EXAMPLE 15.2 (see [**Jon95**, p. 489]). As an important special case, take $U = G$, $K = L$, and $S \leq \mathrm{Stab}_{\mathrm{Aut}(G)}(L)$, a subgroup of $\mathrm{Aut}(G)$ that leaves each L_i invariant. Then $C_S(G)$ is trivial, hence $|\mathrm{Hom}_L(g, G)| \geq |S|$ if $\mathrm{Epi}_L(g, G) \neq \emptyset$. In this case, the proof of the lemma expresses the fact that S acts fixed point freely on $\mathrm{Epi}_L(g, G)$, so $|\mathrm{Epi}_L(g, G)|$ is a multiple of $|S|$.

In particular, for $S = \mathrm{Inn}(G)$, the group of inner automorphisms, the invariance condition is of course satisfied, and $\mathrm{Inn}(G) \cong G/Z(G)$ yields that $|\mathrm{Hom}_L(g, G)| \geq [G{:}Z(G)]$ if $\mathrm{Epi}_L(g, G) \neq \emptyset$.   ◇

EXAMPLE 15.3. For $G \leq U$ and $S = U$, we have of course $Z(G) \subseteq C_S(G)$. If additionally G is maximal and not normal in U then $Z(G) = C_S(G)$, since $C_S(G) \subseteq N_S(G) = G$ implies $C_S(G) = C_G(G) = Z(G)$. In this situation, Lemma 15.1 yields $\mathrm{Epi}_L(g, G) = \emptyset$ if $|\mathrm{Hom}_K(g, U)| < [U{:}Z(G)]$.   ◇

EXAMPLE 15.4. Consider the class structures $C = (2A, 3A, 7A)$ and $C' = (2A, 3A, 7B)$ of the group $G = L_3(2)$. We have $|\mathrm{Hom}_C(0, G)| = |\mathrm{Hom}_{C'}(0, G)| = 168 = |G|$. Since the unique nontrivial outer automorphism of G swaps the classes 7A and 7B, $\mathrm{Stab}_{\mathrm{Aut}(G)}(L)$ is equal to the group $\mathrm{Inn}(G)$ of inner automorphisms of G, which is isomorphic to G.

As shown in Example 11.6, $\mathrm{Epi}_C(0, G)$ and $\mathrm{Epi}_{C'}(0, G)$ are nonempty, so the proof of Lemma 15.1 yields that $\mathrm{Hom}_C(0, G) = \mathrm{Epi}_C(0, G)$ and $\mathrm{Hom}_{C'}(0, G) = \mathrm{Epi}_{C'}(0, G)$.

Note that the orbits of $\mathrm{Aut}(G)$ on $\mathrm{Epi}(\Gamma, G)$ parametrize the normal subgroups with factor group isomorphic to G in Γ (cf. the discussion in Section 5.2). Thus the kernel in $\Gamma(0; 2, 3, 7)$ of epimorphisms onto $L_3(2)$ is unique.
◇

The following statements are obvious but nevertheless useful.

LEMMA 15.5. *Let $L = (L_1, L_2, \ldots, L_r)$ with L_i a union of G-conjugacy classes of element order m_i.*

(a) *If G has a proper normal subgroup that contains all L_i except at most one then $\mathrm{Epi}_L(0, G) = \emptyset$.*

(b) *If $|G| \equiv 2 \pmod 4$ and at most one of the m_i is even then $\mathrm{Epi}_L(0, G) = \emptyset$.*

(c) *If $|G| = 2m_i$ for an i and at most one of the other periods is even then $\mathrm{Epi}_L(0, G) = \emptyset$.*

Proof. Statement (a) follows from the fact that at least two elements in a generating r-tuple must lie outside a proper normal subgroup if the product of elements in the r-tuple is to be the identity.

Statement (b) is a special case of (a) because the elements of odd order in G form a normal subgroup N of index 2. To see this, consider the regular permutation representation of G. Elements of even order consist of an odd number of cycles of even length, hence do not lie in the alternating group $A_{|G|}$. So N is exactly the intersection of G with $A_{|G|}$.

With the same argument, statement (c) also follows from (a), here G has a cyclic normal subgroup N of index 2.   □

LEMMA 15.6. *If a noncyclic group G contains exactly one involution then G is not $(2, m_2, m_3)$-generated, for arbitrary positive integers m_2, m_3.*

Proof. Let z denote the unique involution in G. If G is an epimorphic image of $\Gamma = \Gamma(0; 2, m_2, m_3)$ then so is $G/\langle z \rangle$, and mapping the first elliptic generator of Γ to the identity in $G/\langle z \rangle$ means that the images of the other two generators are mutually inverse. So $G/\langle z \rangle$ is cyclic and hence G is abelian. At least one of m_2, m_3 is even. Let $\sigma \in G$ be the image of the corresponding generator of Γ. Then $G = \langle \sigma \rangle$ because z is a power of σ and thus all generators of Γ are mapped to powers of σ.   □

15.2. Abelian Invariants. Next we utilize *abelian invariants* as a condition on epimorphic images of a group. That is, consider two groups Γ and G, with derived subgroups Γ' and G'. If there is an epimorphism $\Phi \colon \Gamma \to G$ then $\Gamma' \subseteq \Phi^{-1}(G')$. Thus the map $\Phi' \colon \sigma\Gamma' \mapsto \Phi(\sigma)G'$ defines an epimorphism from Γ/Γ' to G/G'.

The existence of the epimorphism Φ' can be decided with Lemma A.2. Note that the kernel of Φ' need not be torsion-free, contrary to the discussion in Section 9.

EXAMPLE 15.7. If all periods of a Fuchsian group Γ with orbit genus 0 are coprime then Γ and hence all its factor groups are perfect.   ◇

EXAMPLE 15.8. The commutator factor group of $\Gamma = \Gamma(0; 2, 5, 5)$ is a cyclic group of order 5, and if G is a factor group of Γ then the commutator factor group of G is either trivial or also cyclic of order 5.

Suppose $|G| = 80$. Then G is solvable, its commutator factor group is of order 5, and G' is a 2-group. So G/G' acts irreducibly on every chief factor of G that is contained in G', and this action cannot be trivial for the chief factor immediately below G' because then G would have a cyclic factor group of order 10. Because the smallest nontrivial representation of a cyclic group of order 5 in characteristic 2 has degree 4, we have $G \cong 2^4 \colon 5$. Furthermore, the kernel of the epimorphism from Γ onto G is torsion-free and has orbit genus 5.   ◇

15.3. Scott's Criterion. A more sophisticated result that allows us to prove that $\mathrm{Epi}_L(0, G) = \emptyset$ in many cases is

THEOREM 15.9 (see [**Sco77**, Theorem 1]). *Suppose the group G is generated by elements x_1, x_2, ... , x_r, with $x_1 x_2 \cdots x_r = 1$. Let G act on the finite-dimensional vector space V over the field F, and let $V^* = \mathrm{Hom}_F(V, F)$ denote the dual module of V. Then*

$$\sum_{i=1}^{r} \dim(V/\mathrm{Fix}_V(x_i)) \geq \dim(V/\mathrm{Fix}_V(G)) + \dim(V^*/\mathrm{Fix}_{V^*}(G)).$$

Proof. Let $C = \{(v_1, v_2, \ldots, v_r) \mid v_i \in (1-x_i)V\}$, and define maps $\delta \colon C \to V$ and $\beta \colon V \to C$ by

$$\delta((v_1, v_2, \ldots, v_r)) = v_1 + x_1 v_2 + \cdots + x_1 x_2 \cdots x_{r-1} v_r$$

and

$$\beta(v) = ((1 - x_1)v, (1 - x_2)v, \ldots, (1 - x_r)v),$$

respectively. Then

$$1 - x_1 x_2 \cdots x_r = 1 - x_1 + x_1(1 - x_2) + \cdots + x_1 x_2 \cdots x_{r-1}(1 - x_r)$$

implies

$$\delta((1 - x_1)v, \ldots, (1 - x_r)v)$$
$$= (1 - x_1)v + x_1(1 - x_2)v + \cdots + x_1 \cdots x_{r-1}(1 - x_r)v = 0,$$

that is, $\beta(V) \subseteq \ker \delta$. We have $\ker \beta = \bigcap_{i=1}^{r} \mathrm{Fix}_V(x_i) = \mathrm{Fix}_V(G)$, thus $\dim \beta(V) = \dim(V/\mathrm{Fix}_V(G))$.

$\delta(C) = \sum_{i=1}^{r}(1 - x_i)V$ is the smallest submodule U of V with trivial action of G on the factor V/U; to see this, note that the action on $V/\delta(C)$ is trivial because $x_i v \equiv v \pmod{\delta(C)}$, and that the condition $x_i v \equiv v \pmod{U}$ implies $(x_i - 1)V \subseteq U$, hence $\delta(C) \subseteq U$.

Consequently, $\mathrm{Fix}_{V^*}(G)$ is the annihilator of $\delta(C)$ in V^*, and thus $\dim \delta(C) = \dim(V^*/\mathrm{Fix}_{V^*}(G))$.

Putting these facts together, we get

$$
\sum_{i=1}^{r} \dim(V/\mathrm{Fix}_V(x_i)) = \dim C
$$

$$
\begin{aligned}
&= \dim(C/\ker \delta) + \dim(\ker \delta/\beta(V)) + \dim \beta(V) \\
&\geq \dim \delta(C) + \dim \beta(V) \\
&= \dim(V^*/\mathrm{Fix}_{V^*}(G)) + \dim(V/\mathrm{Fix}_V(G)).
\end{aligned}
$$

$\square$

From this theorem, we get immediately a character theoretic nongeneration criterion.

COROLLARY 15.10. *Let χ be a character of the finite group G such that*

$$
\sum_{i=1}^{r} \left(\chi(1) - [\chi_{\langle x_i \rangle}, 1_{\langle x_i \rangle}] \right) < 2 \left(\chi(1) - [\chi, 1_G] \right)
$$

for elements x_1, x_2, ..., x_r of G that satisfy $x_1 x_2 \cdots x_r = 1$. Then x_1, x_2, ..., x_r generate a proper subgroup of G.

If the condition of the corollary holds for χ then it holds for one of its irreducible constituents. Thus it is sufficient to check the irreducible characters of G, if they are known.

EXAMPLE 15.11. We show that the Mathieu group M_{22} of order $443\,520$ is not an epimorphic image of $\Gamma(0; 2, 3, 7)$. For that, we consider the irreducible character $\chi = \chi_2$ of degree 21 of M_{22}. Its values on elements of orders 2, 3, and 7 are $\chi(2\mathrm{A}) = 5$, $\chi(3\mathrm{A}) = 5$, $\chi(7\mathrm{A}) = \chi(7\mathrm{B}) = 0$, respectively. Thus the fixed spaces have dimensions 13, 9, and 3, and since $(21-13)+(21-9)+(21-3) = 38 < 2 \cdot 21$ the theorem yields nongeneration. ◇

EXAMPLE 15.12. The signatures $(0; 2, 2, n)$, $(0; 2, 3, 3)$, $(0; 2, 3, 4)$, $(0; 2, 3, 5)$ of the finite polyhedral groups D_{2n}, A_4, S_4, and A_5 that have been mentioned already in Section 3.4 need not be considered in the context of automorphism groups of compact Riemann surfaces of genus at least 2. And for questions whether a given class structure of a group generates this group, they can be discarded a priori if the group has none of the types listed above.

But these signatures are also excluded by Corollary 15.10. That is, for a group G, let H be the polyhedral subgroup in question, and choose the character $\chi = 1_H^G$ of the action of G on the cosets $\Omega = G/H$ of H. Observe that the dimension of the fixed space of an element σ in a representation with character χ equals the number of orbits of σ on Ω. If σ has order m and f fixed points then it has $f + (|\Omega| - f)/m$ orbits. Moreover, an element in H fixes at least one point. So the condition of Corollary 15.10 is satisfied if

$$3|\Omega| - \sum_{i=1}^{3} (1 + (|\Omega| - 1)/m_i) < 2(|\Omega| - 1)$$

holds. But this is the case if $|\Omega| \geq 1$, i.e., if H is a proper subgroup of G, because $\sum_{i=1}^{3}(1/m_i) > 1$.   ◇

EXAMPLE 15.13. Lemma 15.5 (a) is also implied by Corollary 15.10. Suppose that N is a proper normal subgroup of G that contains at least $r - 1$ of the elements $x_1, x_2, \ldots, x_r$. Set $\chi = 1_N^G$, then $[\chi_{\langle n \rangle}, 1_{\langle n \rangle}] = \chi(1)$ for $n \in N$, by the top formula on page 24. Thus the value of left hand side of the inequality in Corollary 15.10 is at most $\chi(1) - 1$, and the value of the right hand side is $2(\chi(1) - 1)$.   ◇

15.4. Utilizing Eichler's Trace Formula. Theorem 12.1 together with Lemmas 3.13 and 11.5 can be interpreted as a recipe to assign a character to each generating r-tuple for a group. From a more general viewpoint, we get a criterion of nongeneration.

LEMMA 15.14. *Let g be a nonnegative integer, and $C = (C_1, C_2, \ldots, C_r)$ be a class structure of the group G, with C_i consisting of elements of order m_i. Define the class function ψ of G by*

$$\psi(1) = 1 + (g - 1)|G| + \frac{|G|}{2} \cdot \sum_{i=1}^{r} (1 - 1/m_i)$$

and

$$\psi(\sigma) = 1 + \sum_{u \in I(m)} a_u(\sigma) \frac{\zeta_m^u}{1 - \zeta_m^u}$$

for $\sigma \in G^\times$ of order m, where

$$a_u(\sigma) = |C_G(\sigma)| \sum_{1 \leq i \leq r, \sigma \in C_i} \frac{1}{m_i}.$$

(a) *If $\mathrm{Epi}_C(g, G) \neq \emptyset$ then ψ is a character of G.*
(b) *If $\mathrm{Hom}_C(g, G) \neq \emptyset$ then ψ is a virtual character of G.*

Proof. If $\mathrm{Epi}_C(g, G) \neq \emptyset$ then we may choose a surface kernel epimorphism $\Phi \colon \Gamma(g; m_1, m_2, \ldots, m_r) \to G$, and G acts as a group of automorphisms on

$X = \mathcal{U}/\ker(\Phi)$ of genus $\psi(1)$ such that $|\mathrm{Fix}_{X,u}(\sigma)| = a_u(\sigma)$ for all $\sigma \in G^{\times}$ and $u \in I(|\sigma|)$. So ψ is the character of the action of G on $\mathcal{H}^1(X)$ by Theorem 12.1.

If $\mathrm{Hom}_C(g, G) \neq \emptyset$ then $\mathrm{Epi}_C(g + k, G) \neq \emptyset$ if k is large enough, and the class function associated to an epimorphism in this set is $\psi + k \cdot \rho_G$, which is a character by part (a). $\qquad\square$

In practice, this criterion turns out to be not very strong. In particular, no example is known for which Lemma 15.14 yields nongeneration but Corollary 15.10 does not.

EXAMPLE 15.15. The group $G = 2.A_5$ is not $(3, 3, 4)$-generated because the class function associated to the unique class structure of G with these element orders is $-\chi_4 + \chi_6 + \chi_7 + \chi_9$. $\qquad\diamond$

15.5. Proving Generation. Now we change the perspective, and look for criteria that guarantee generation.

LEMMA 15.16. *Let $C = (C_1, C_2, \ldots, C_r)$ be a class structure of G, such that each maximal cyclic subgroup of G has a generator in one of the C_i. Then $\mathrm{Hom}_C(g, G) = \mathrm{Epi}_C(g, G)$.*

Proof. Suppose $H \leq G$ is the image of an element in $\mathrm{Hom}_C(g, G)$. By our assumptions, H contains elements from all conjugacy classes of G, so all values of 1_H^G are strictly positive, by the top formula on page 24. Because $|G| = |G| \cdot [1_H^G, 1_H] = \sum_{\sigma \in G} 1_H^G(\sigma)$, this means that $1_H^G = 1_G$. In other words, $H = G$. $\qquad\square$

In some cases, additional information about (maximal) subgroups of G in terms of characters can be used to prove that elements from prescribed conjugacy classes generate G.

LEMMA 15.17. *Let $C = (C_1, C_2, \ldots, C_r)$ be a class structure of G, and choose $\sigma_i \in C_i$, for $1 \leq i \leq r$. If $\prod_{i=1}^r \pi(\sigma_i) = 0$ for all permutation characters $\pi = 1_H^G$ of G, where H runs over all (conjugacy classes of) maximal subgroups H of G, then $\mathrm{Hom}_C(g, G) = \mathrm{Epi}_C(g, G)$.*

Proof. By the top formula on page 24, $1_H^G(\sigma) = 0$ if and only if H contains no element of the conjugacy class of σ. So the condition of the lemma means that no proper subgroup of G contains elements of all classes C_1, C_2, $\ldots$, C_r. $\qquad\square$

EXAMPLE 15.18. Consider the class structure $C = (2\mathrm{C}, 4\mathrm{A}, 5\mathrm{E})$ of $G = S_4(4)$. We have $|\mathrm{Hom}_C(0, G)| = 3\,916\,800 = |\mathrm{Aut}(G)|$. As one can see from the permutation characters of the maximal subgroups of G listed in [**CCN$^+$85**,

p. 44], and from the possible class fusions of the subgroup of type S_6 in G, the first four classes of maximal subgroups do not contain 5E elements, and the remaining classes of maximal subgroups do not contain 4A elements. Thus $\mathrm{Epi}_C(0, G) = \mathrm{Hom}_C(0, G)$ by Lemma 15.17, hence $S_4(4)$ is $(2, 4, 5)$-generated.

◇

If not only the primitive permutation characters of G but the character tables of maximal subgroups of G are available, we can apply a stronger criterion.

LEMMA 15.19. *Fix a class structure* $C = (C_1, C_2, \ldots, C_r)$ *of* G. *For a subgroup* H *of* G, *let* $C \cap H = (C_1 \cap H, C_2 \cap H, \ldots, C_r \cap H)$. *Then*

$$|\mathrm{Epi}_C(g, G)| \geq |\mathrm{Hom}_C(g, G)| - \sum_{\substack{H < G \\ \text{maximal}}} |\mathrm{Hom}_{C \cap H}(g, H)|.$$

Proof. We have

$$\mathrm{Epi}_C(g, G) = \mathrm{Hom}_C(g, G) \Big\backslash \bigcup_{H < G} \mathrm{Hom}_{C \cap H}(g, H),$$

where the union of subgroups H may clearly be restricted to the maximal subgroups of G.   □

Lemma 15.17 describes just the special case of the above lemma that at least one component of $C \cap H$ is empty for each proper subgroup H of G. Note that $C \cap H$ is in general not a class structure of H but describes a union of class structures.

The right hand side of the formula in Lemma 15.19 can be computed from the character tables of G and its maximal subgroups, provided the fusions of conjugacy classes of the maximal subgroups into G are known. Of course one needs to consider only representatives of conjugacy classes of maximal subgroups; the value $|\mathrm{Hom}_{C \cap H}(g, H)|$ for each such subgroup H must then be multiplied by the length of the conjugacy class of H, which is $[G{:}H]$ if H is not normal in G, and 1 otherwise.

EXAMPLE 15.20. Take $G = M_{12}$, the Mathieu group of order 95 040. We want to show that G is $(2, 3, 10)$-generated. More precisely, let C be the class structure (2B, 3B, 10A) of G, then $|\mathrm{Hom}_C(0, G)| = 190\,080 = |\mathrm{Aut}(G)|$. Exactly one class of maximal subgroups in G contains elements in all three classes of C, see [CCN+85, p. 33], namely the class of H with $H \cong 2 \times S_5$. Because $|\mathrm{Hom}_{C \cap H}(0, H)| = 0$, we get $\mathrm{Epi}_C(0, G) = \mathrm{Hom}_C(0, G)$.   ◇

We can combine the idea of Lemma 15.19 with Lemma 15.1 to get a criterion of *nongeneration*. This is a standard argument for example in [CWW92] but stated there only implicitly, probably because the formulation is rather technical. The idea is to exploit, for an arbitrary proper subgroup H of G, the relation

$$\text{Epi}_C(g, G) \subseteq \text{Hom}_C(g, G) \setminus \text{Hom}_{C \cap H}(g, H),$$

in the situation of Lemma 15.19.

LEMMA 15.21. *Let G, H, C, $C \cap H$ be as in Lemma 15.19, and* $\text{Stab}_{\text{Aut}(G)}(C)$ *as in Example 15.2. For $1 \leq i \leq r$, let $C_i \cap H = \biguplus_{j \in J_i} K(i, j)$, a disjoint union of conjugacy classes $K(i, j)$ in H. Set $L(i, j) = (L(i, j)_1, L(i, j)_2, \ldots, L(i, j)_r)$, where $L(i, j)_i = K(i, j)$ and $L(i, j)_k = C_k \cap H$ for $k \neq i$. If*

$$|\text{Hom}_C(g, G)| - |\text{Stab}_{\text{Aut}(G)}(C)|$$

$$< \max \left\{ |\text{Hom}_{L(i,j)}(g, H)| \cdot \frac{|C_i|}{|K(i, j)|} \,\middle|\, 1 \leq i \leq r, j \in J_i \right\}$$

then $\text{Epi}_C(g, G) = \emptyset$.

Proof. Fix $i \in \{1, 2, \ldots, r\}$, $j \in J_i$, and choose an element $z \in K(i, j)$. We consider only homomorphisms that map the i-th elliptic generator to z, and denote the corresponding sets by $\text{Hom}_C^{i,z}(g, G)$, $\text{Epi}_C^{i,z}(g, G)$, and $\text{Hom}_{L(i,j)}^{i,z}(g, H)$. Then we have

$$\text{Epi}_C^{i,z}(g, G) \subseteq \text{Hom}_C^{i,z}(g, G) \setminus \text{Hom}_{L(i,j)}^{i,z}(g, H),$$

and thus

$$\frac{|\text{Epi}_C(g, G)|}{|C_i|} \leq \frac{|\text{Hom}_C(g, G)|}{|C_i|} - \frac{|\text{Hom}_{L(i,j)}(g, H)|}{|K(i, j)|}.$$

Now we multiply this inequality by $|C_i|$, and get

$$|\text{Epi}_C(g, G)| \leq |\text{Hom}_C(g, G)| - |\text{Hom}_{L(i,j)}(g, H)| \cdot \frac{|C_i|}{|K(i, j)|},$$

for all i and j, by the fact that we may choose z in any of the classes $K(i, j)$. By our assumption, this implies $|\text{Epi}_C(g, G)| < |\text{Stab}_{\text{Aut}(G)}(C)|$, and the claim follows from Lemma 15.1 or Example 15.2, with $U = G$, $K = L = C$, and $S = \text{Stab}_{\text{Aut}(G)}(C)$. □

REMARK 15.22.   (a) The relation mentioned before Lemma 15.21 yields

$$|\text{Epi}_C(g, G)| \leq |\text{Hom}_C(g, G)| - |\text{Hom}_{C \cap H}(g, H)|,$$

which is in general weaker than Lemma 15.21. That is, for each i we have

$$
\begin{aligned}
|\mathrm{Hom}_{C \cap H}(g, H)| &= \sum_{j \in J_i} |\mathrm{Hom}_{L(i,j)}(g, H)| \\
&= \sum_{j \in J_i} |K(i,j)| \cdot \frac{|\mathrm{Hom}_{L(i,j)}(g, H)|}{|K(i,j)|} \\
&\leq \sum_{j \in J_i} |K(i,j)| \cdot \max\left\{ \frac{|\mathrm{Hom}_{L(i,k)}(g, H)|}{|K(i,k)|} \,\middle|\, k \in J_i \right\} \\
&= |C_i \cap H| \cdot \max\left\{ \frac{|\mathrm{Hom}_{L(i,k)}(g, H)|}{|K(i,k)|} \,\middle|\, k \in J_i \right\} \\
&\leq |C_i| \cdot \max\left\{ \frac{|\mathrm{Hom}_{L(i,k)}(g, H)|}{|K(i,k)|} \,\middle|\, k \in J_i \right\} \\
&= \max\left\{ |\mathrm{Hom}_{L(i,k)}(g, H)| \cdot \frac{|C_i|}{|K(i,k)|} \,\middle|\, k \in J_i \right\}.
\end{aligned}
$$

(b) Analogously one can try to improve Lemma 15.19 by considering only those homomorphisms with one fixed image z in one of the classes C_i. But the argument of Lemma 15.21 applies only for *one* fixed maximal subgroup H, so one can replace at most the summand for *one* maximal subgroup H involved by a smaller one; already for the conjugates of H different from H, in general no such improvement is possible.

EXAMPLE 15.23. Take $G = U_3(3)$, the simple group of order 6048. If C is one of the class structures $(2A, 3A, 7A)$ and $(2A, 3A, 7B)$ then $\mathrm{Hom}_C(0, G) = \emptyset$, and for the class structures $C = (2A, 3B, 7A)$ and $C' = (2A, 3B, 7B)$, we have $|\mathrm{Hom}_C(0, G)| = |\mathrm{Hom}_{C'}(0, G)| = 6048 = |G|$. There is a class of maximal subgroups $H \cong L_3(2)$ in G, see [**CCN+85**, p. 14], for which $|\mathrm{Hom}_{C \cap H}(0, H)| = |\mathrm{Hom}_{C' \cap H}(0, H)| = 168 = |H|$. So $U_3(3)$ is not $(2, 3, 7)$-generated.

Moreover, the $(2, 3, 7)$-subgroups of G are exactly the conjugates of H, because each of the homomorphisms to H is surjective by Example 15.4, and the number of conjugates of H in G is $|G|/|H|$.   ◇

EXAMPLE 15.24 (cf. [**CWW92**]). We show that the sporadic simple Janko group $G = J_3$ of order 50 232 960 is not $(2, 3, 8)$-generated. There are the two class structures $C_1 = (2A, 3A, 8A)$ and $C_2 = (2A, 3B, 8A)$ to consider. C_2 is ruled out by Corollary 15.10, applied to an irreducible degree 85 character of G. For excluding C_1 also, we first compute $|\mathrm{Hom}_{C_1}(0, G)| = 226\,048\,320 = \frac{9}{2}|G|$ and then look at a maximal subgroup H of isomorphism type $(3 \times A_6) : 2_2$ in G (see [**CCN+85**, p. 82]). We have

$$
\mathrm{Hom}_{C \cap H}(0, H) = \biguplus_{1 \leq j \leq 4} \mathrm{Hom}_{K_j}(0, H)
$$

for class structures $K_1 = (2a, 3a, 8a)$, $K_2 = (2a, 3b, 8a)$, $K_3 = (2a, 3a, 8b)$, $K_4 = (2a, 3b, 8b)$ of H, each with cardinality $|\mathrm{Hom}_{K_j}(0, H)| = |H| = 2\,160$. The lengths of the H-classes are 108, 80, and 270, for the elements of order 2, 3, and 8, respectively. Trying to maximize the right hand side of the inequality in Lemma 15.21, we see that the maximum is attained for $L(3, 1) = (2a, 3a \cup 3b, 8a) = K_1 \cup K_2$, with value $2|G|$. Because of $|\mathrm{Stab}_{\mathrm{Aut}(G)}(C_1)| = |\mathrm{Aut}(G)| = 2|G|$, this is –unfortunately– smaller than $\frac{9}{2}|G| - 2|G| = \frac{5}{2}|G|$, so Lemma 15.21 does *not* exclude C_1. (Note that the argument in the proof of [**CWW92**, Proposition 1] is not correct.)

But we can apply a slightly modified version of the lemma, namely, if H is maximal but not normal in G then the right hand side of the inequality can be replaced by

$$\max\left\{ \sum_{j \in J_i} |\mathrm{Epi}_{L(i,j)}(g, H)| \cdot \frac{|C_i|}{|K(i,j)|} \,\middle|\, 1 \le i \le r \right\}.$$

To see this, choose for each $j \in J_i$ elements $z_j \in K(i, j)$ and $\sigma_j \in G$ with $z_j^{\sigma_j} = z$. Then each of the groups H^{σ_j} contains z, and the H^{σ_j} are pairwise distinct. The latter follows from the fact that $H^{\sigma_1} = H^{\sigma_2}$ would imply $\sigma_1 \sigma_2^{-1} \in N_G(H)$, which is equal to H by our assumption, and thus $\sigma_1 \sigma_2^{-1}$ would be an element in H that conjugates z_1 to z_2, contrary to our choice of the z_j in different H-classes. Now the claim follows from

$$\biguplus_{j \in J_i} \mathrm{Epi}^{i,z}_{L(i,j)^{\sigma_j}}(g, H^{\sigma_j}) \subseteq \mathrm{Hom}^{i,z}_C(g, G) \setminus \mathrm{Epi}^{i,z}_C(g, G).$$

So we are done if we have shown that $\mathrm{Hom}_{K_j}(0, H) = \mathrm{Epi}_{K_j}(0, H)$ holds for $1 \le j \le 4$. The character table of H suffices for this. We compute, with the help of GAP, a list of all H-characters that have certain properties of permutation characters (see [**BP98**]); there are 59 such characters, only the trivial character is nonzero on each of $2a, 3a \cup 3b, 8a \cup 8b$, so Lemma 15.17 can be applied. ◇

Finally, we state a criterion that relates two class structures.

LEMMA 15.25 (see [**CWW92**, Lemma 2]). *Let G be a finite simple group, $C = (\sigma^G, \tau^G, \kappa^G)$, $C' = (\tau^G, \tau^G, (\kappa^2)^G)$, and $C'' = (\kappa^G, \kappa^G, (\tau^2)^G)$ class structures of G, with σ of order 2. If $\mathrm{Epi}_C(0, G) \ne \emptyset$ then $\mathrm{Epi}_{C'}(0, G) \ne \emptyset$ and $\mathrm{Epi}_{C''}(0, G) \ne \emptyset$.*

Proof. Suppose $G = \langle \sigma, \tau \rangle$, with $\kappa = (\sigma\tau)^{-1}$. Then $\langle \tau^\sigma, \tau \rangle$ is a non-trivial normal subgroup in G, hence equal to G. Now note that $(\tau^\sigma \tau)^{-1} = (\sigma\tau\sigma\tau)^{-1} = \kappa^2$. The second case follows because with $\sigma\tau\kappa$, $\sigma^\tau \kappa\tau$ is also the identity. □

EXAMPLE 15.26. Let $G = M_{12}$, the Mathieu group of order 95 040. The class structure $(2A, 3A, 11A)$ generates G because no proper subgroup of G contains elements in all three classes. Thus Lemma 15.25 yields that the class structure $(3A, 3A, 11B)$ also generates G; note that the class $11B$ consists of the squares of the elements in the class $11A$. Analogously, the class $6B$ squares to $3A$, and we conclude that with $(2A, 6B, 8A)$, $(3A, 8A, 8A)$ also generates G. Applying Lemma 15.25 in the other direction, we see that $(2A, 4A, 10A)$ does *not* generate G because $(2B, 10A, 10A)$ does not generate G and $4A$ squares to $2B$. These cases cannot be decided with the criteria we had discussed before Lemma 15.25. ⋄

15.6. Brute Force Check of Generation. For a given g and a given class structure C of G, we cannot expect that the above criteria always provide an answer to the question whether $\mathrm{Epi}_C(g, G)$ is empty. In this case, we may loop over the set of G-orbits in $\mathrm{Hom}_C(g, G)$, and check whether one of them describes epimorphisms. (Of course G must admit the necessary computations if this "brute force test" is to be applied.) An advantage of this approach, compared with the criteria presented before, is that it is constructive, i.e., one gets the epimorphisms themselves and not only the guarantee that they exist. Orbit representatives can be computed using

LEMMA 15.27. *Let* $C = (C_1, C_2, \ldots, C_r)$ *be a class structure of* G, *and* $\sigma_i \in C_i$ *for* $1 \leq i \leq r$. *A system of orbit representatives of* $C_1 \times C_2 \times \cdots \times C_r$ *under the action of* G *via* $(\sigma_1, \sigma_2, \ldots, \sigma_r)^\alpha = (\sigma_1^\alpha, \sigma_2^\alpha, \ldots, \sigma_r^\alpha)$ *is given by*

$$\{(\sigma_1, \sigma_2^{\beta_2}, \ldots, \sigma_r^{\beta_r}) \mid \beta_i \in R(\beta_2, \ldots, \beta_{i-1}) \quad \text{for} \quad 2 \leq i \leq r\},$$

where the set $R(\beta_2, \ldots, \beta_{i-1})$ *is defined iteratively as system of representatives of the double cosets* $C_G(\sigma_i) \backslash G / C_G(\sigma_1, \sigma_2^{\beta_2}, \ldots, \sigma_{i-1}^{\beta_{i-1}})$.

Proof. By induction on r, we may assume that a given r-tuple is G-conjugate to $(\sigma_1, \sigma_2^{\beta_2}, \ldots, \sigma_{r-1}^{\beta_{r-1}}, \sigma_r^\beta)$ for suitable $\beta_i \in R(\beta_2, \ldots, \beta_{i-1})$ and $\beta \in G$. Now we may conjugate this r-tuple by any element in $C_G(\sigma_1, \sigma_2^{\beta_2}, \ldots, \sigma_{r-1}^{\beta_{r-1}})$ without changing the first $r-1$ components. Hence we achieve a representative in $R(\beta_2, \ldots, \beta_{r-1})$ as conjugating element in the r-th component. □

Note that Lemma 15.27 describes an algorithm to loop over the G-orbits on $\mathrm{Hom}_C(g, G)$ for $g = 0$, and that the proof of Theorem 14.1 shows how to generalize it to the case $g > 0$, by taking disjoint unions.

Furthermore, if the action of $\mathrm{Stab}_{\mathrm{Aut}(G)}(C)$ on G is known (see Lemma 15.1) then we can use the action of this group instead of that of $G/Z(G)$.

EXAMPLE 15.28. As an application of Lemma 15.27, we consider a problem similar to that of whether a group is generated by a given class structure.

That is, we want to know whether the orthogonal group $G = O_8^+(3) \cong PSO_8^+(3)'$ is generated by two elements σ, τ in the conjugacy classes 13A, 3A, respectively.

First we try to answer the question with the help of the character theoretic criteria. The conjugacy class of the product $\sigma\tau$ is not prescribed, that is, we consider all class structures $C = (13A, 3A, C_3)$ of G for which $\mathrm{Hom}_C(0, G)$ is nonempty. According to the list of maximal subgroups of G in [CCN$^+$85, p. 140], only maximal subgroups of the types $O_7(3)$ and $3^6 : L_4(3)$ contain elements of order 13, and their character tables are available in GAP, as is the table of G. From the fusions of conjugacy classes of these maximal subgroups in G, we obtain that all class structures under consideration may contain triples that generate proper subgroups of G, so Lemma 15.17 cannot be applied. Also, Lemma 15.19 does not answer our question. So we try the "brute force test". Lemma 15.27 tells us how to distribute the triples that belong to each class structure into orbits under the action of G. Since the third class of the class structure is not prescribed, we can use a variant of this lemma. Let $K = N_G(\tau) \cong 3_+^{1+8} : 2(A_4 \times A_4 \times A_4).2$, $H = \langle\sigma\rangle$, and R be a set of representatives of K-H-double cosets in G. Because $\langle\sigma, \tau^{kr\sigma^i}\rangle = \langle\sigma, \tau^c\rangle^{\sigma^i}$ for $k \in K$ and $r \in R$, either all elements in a double coset generate G, or none of them does. (But the conjugacy class of the product of the two generators is not invariant.) K is a maximal subgroup of index 36 400 in G, and $|K \cap H| = 1$ yields $|R| = \frac{|G|}{|K||H|} = 36\,400/13 = 2\,800$.

Suppose G is given in its permutation representation of the cosets of K, i.e., in the unique primitive permutation representation on 36 400 points. For $1 \le i \le 36\,400$, let $r_i \in G$ map 1 to i. The double cosets Kr_iH, Kr_jH are equal if and only if i and j lie in the same orbit under H, so we may choose a set I of H-orbit representatives, and $R = \{r_i \mid i \in I\}$. In order to test whether $\langle\sigma, \tau^r\rangle = G$ for $r \in R$, it is sufficient to check whether $\langle\sigma, \tau^r\rangle$ acts transitively on the 36 400 points. This is because the orders of the $O_7(3)$ and $3^6 : L_4(3)$ are not divisible by 36 400, and hence no subgroup of these groups can act transitively on 36 400 points.

For the construction of the permutation representation of G, we use the computer systems GAP and MeatAxe (see [S$^+$94, Rin94]). Starting with the 8-dimensional matrix representation of $SO_8^+(3)'$ over the field with 3 elements, we find a permutation representation on 1 080 points in the action on the 1-dimensional subspaces of the natural module. This permutation module, reduced modulo 2, has a composition factor of dimension 298. In the action of K on this module, there is a nontrivial fixed vector. By the maximality of K in G, the orbit of this vector under G has length 36 400, which yields the desired representation.

Finally, we make the $2\,800$ transitivity checks. Exactly 432 candidates are transitive, so G can be generated by an element in the class 13A and one in 3A. Moreover, the probability that a randomly chosen pair of elements in the classes 13A and 3A of G generates the whole group is $432/2\,800$, this is a chance of about 15 percent.   ◇

EXAMPLE 15.29. The G be the Mathieu group M_{11} of order $7\,920$. We apply our criteria of generation and nongeneration to all class structures $C = (C_1, C_2, C_3)$ of G. Since G has 10 conjugacy classes and the class of the identity need not be considered, there are $\binom{9+3-1}{3} = 165$ triples C to check; note that only *unordered* triples need to be considered by Remark 14.2. Table 5 lists the results of the character theoretic criteria. Only for one class structure out of 130 possibilities, namely $(3A, 5A, 5A)$, do these criteria admit no decision whether it generates G. With the procedure described in Lemma 15.27, we can show that M_{11} is not $(3, 5, 5)$-generated.   ◇

Criterion	Nec. generating	Poss. generating
Lemma 15.1		155
Corollary 15.10		134
Lemma 15.14		134
Lemma 15.21		145
Lemma 15.17	47	
Lemma 15.19	129	
All above together	129	130

TABLE 5. (Non)Generating Triples for M_{11}

15.7. Orbits on Class Structures. Analogously to the action of the group $\mathrm{Gal}(\mathrm{Irr}(G))$ (see Section 6) on the (irreducible) characters of G, we can define an action of this group on the conjugacy classes of G and hence on the class structures. That is, for a class structure $C = (C_1, C_2, \ldots, C_r)$ of G and $*k \in \mathrm{Gal}(\mathrm{Irr}(G))$, let C_i^{*k} denote the class that contains the k-th powers of the elements in C_i, and $C^{*k} = (C_1^{*k}, C_2^{*k}, \ldots, C_r^{*k})$.

LEMMA 15.30. *For each class structure C of G and $*k \in \mathrm{Gal}(\mathrm{Irr}(G))$, we have $|\mathrm{Hom}_C(g, G)| = |\mathrm{Hom}_{C^{*k}}(g, G)|$ and $|\mathrm{Epi}_C(g, G)| = |\mathrm{Epi}_{C^{*k}}(g, G)|$.*

Proof. The first statement follows directly from the formula in Theorem 14.1, since we may view the action of $*k$ on the classes by the action on the irreducible characters, which means just a reordering of the summands.

The second statement follows from the first by induction. To see this, observe that

$$\mathrm{Epi}_C(g, G) = \mathrm{Hom}_C(g, G) \setminus \biguplus_{H < G} \mathrm{Epi}_{C \cap H}(g, H),$$

thus the desired equality is equivalent to $|\mathrm{Epi}_{C \cap H}(g, H)| = |\mathrm{Epi}_{C*^k \cap H}(g, H)|$ for all proper subgroups H of G, and this holds because $\mathrm{Epi}_{C*^k \cap H}(g, H)$ is the union of the images of the class structures in $\mathrm{Epi}_{C \cap H}(g, H)$ under $*k$. $\square$

In practice, Lemma 15.30 can of course be used to reduce the number of tests when several class structures are checked for generating a group. Another application is shown in the following examples.

EXAMPLE 15.31. Let $G = L_2(13)$, see [**CCN$^+$85**, p. 8], and $\Gamma = \Gamma(0; 2, 3, 7)$. No proper subgroup of G contains elements of orders 2, 3, and 7, thus we have $\mathrm{Epi}(\Gamma, G) = \mathrm{Hom}(\Gamma, G)$. Because $|\mathrm{Hom}(\Gamma, G)| = 6|G| = 3|\mathrm{Aut}(G)|$ and because the normal subgroups N in Γ with $\Gamma/N \cong G$ are in bijection with the orbits of $\mathrm{Aut}(G)$ on $\mathrm{Epi}(\Gamma, G)$, Γ has exactly three (torsion-free) normal subgroups with factor group isomorphic to G.

Each of these normal subgroups corresponds to a Riemann surface with (full) automorphism group G (see Section 5.2). The genus of these surfaces is 14, and since χ_9 is the only character of degree 14 of G that satisfies the condition of Corollary 12.3, χ_9 comes from all three Riemann surfaces. $\diamond$

EXAMPLE 15.32 (see [**Sah69**, Proposition 2.7]). Let $G = J_1$, Janko's sporadic simple group of order 175 560. This group has unique conjugacy classes of element orders 2, 3, and 7. No proper subgroup of G is $(2, 3, 7)$-generated, so $\mathrm{Epi}(\Gamma, G) = \mathrm{Hom}(\Gamma, G)$ holds for $\Gamma = \Gamma(0; 2, 3, 7)$. The cardinality of the sets is $7 \cdot |G|$. Now $\mathrm{Aut}(G) = G$ implies that Γ has exactly seven torsion-free normal subgroups N with $\Gamma/N \cong G$. $\diamond$

EXAMPLE 15.33. As in Example 15.29, take $G = M_{11}$, and consider the class structure $C = (2A, 4A, 11A)$. We have $|\mathrm{Epi}_C(0, G)| = 7\,920 = |G|$, so the group $\Gamma = \Gamma(0; 2, 4, 11)$ has exactly one torsion-free normal subgroup K such that $\Gamma/K \cong G$ and such that the image of the elliptic generator of order 11 lies in the conjugacy class 11A. Moreover, since the outer automorphism group of M_{11} is trivial, Γ has exactly *two* torsion-free normal subgroups with factor group isomorphic to G. The first one, K, corresponds to $|\mathrm{Epi}_C(0, G)|$, the other one, $\overline{K}$, belongs to the Galois conjugate class structure $(2A, 4A, 11B)$. The classes of element order 11 in G are not real, so the characters that come from the Riemann surfaces $\mathcal{U}/K$ and $\mathcal{U}/\overline{K}$ are also not real, in fact they are mutually complex conjugate. $\diamond$

16. The Strong Symmetric Genus of a Group

By Corollary 3.15, any finite group can be realized as a group of automorphisms of a compact Riemann surface, and it is a natural question what is the smallest genus g of a compact Riemann surface with a group of automorphisms isomorphic to a given finite group G. Following [**Tuc83**], this number g is called the *strong symmetric genus* of G.

The groups of genus 0 and 1 have been classified in Section 3.4, so let the strong symmetric genus g of G be at least 2. In this case, the inequality of Theorem 3.17 can be read as $g \geq 1 + |G|/84$, with equality if and only if G is a Hurwitz group.

The strong symmetric genus is known for alternating and symmetric groups (see [**Con85**]), for abelian groups (see [**Mac65**]), and for most of the sporadic simple groups (see [**CWW92**] for an overview). Furthermore, the Hurwitz groups of order up to 10^6 have been classified in [**Con87**]. A survey of techniques to decide whether or not a class structure generates a group is given in [**Con91**], where the strong symmetric genus of the Mathieu groups is computed using exhaustive searching as in Lemma 15.27 and alternatively using the enumeration of classes of subgroups of low index in a finitely presented group.

The idea of how to proceed in the computation of the strong symmetric genus of a given group G is straightforward. First, each class structure that is known to generate G yields an upper bound on the strong symmetric genus of G. If G is given by generators, for example as a permutation group or a matrix group, then this provides such a generating class structure. Second, once we have an upper bound we must check only those finitely many signatures that may lead to a smaller genus.

We can also obtain the results for several of the sporadic simple groups (among those, for all Mathieu groups) with the (nonconstructive) character theoretic tools developed in Section 15. Note that the character tables of most maximal subgroups of the sporadic simple groups are easily accessible in GAP. Table 6 lists the numbers of conjugacy classes (Cl.), the numbers of triples to consider (Tr.), the numbers of possibly (Poss.) and necessarily (Nec.) generating triples according to the character theoretic criteria of Section 15, the smallest genus that was proved with the criteria (g), a generating triple (C_1, C_2, C_3) for which this genus is attained, and in the last column a plus sign if the criteria suffice to prove that the genus shown is in fact the strong symmetric genus and a minus sign otherwise.

We checked all unordered triples for the groups listed in Table 6, overgroups used were $HS.2$ for HS, Co_3 for McL, $He.2$ for He, and Co_1 for Co_2. Only

G	Cl.	Tr.	Poss.	Nec.	g	(C_1, C_2, C_3)	
M_{11}	10	165	130	129	631	(2A, 4A, 11A)	+
M_{12}	15	560	394	382	3 169	(2B, 3B, 10A)	+
J_1	15	560	541	537	2 091	(2A, 3A, 7A)	+
M_{22}	12	286	241	234	34 849	(2A, 5A, 7A)	+
J_2	21	1 540	1 149	1 120	7 201	(2B, 3B, 7A)	+
M_{23}	17	816	753	739	1 053 361	(2A, 4A, 23A)	+
HS	24	2 300	1 957	1 878	1 680 001	(2B, 3A, 11A)	+
J_3	21	1 540	1 487	1 476	1 255 825	(2A, 4A, 5A)	+
M_{24}	26	2 925	2 630	2 561	10 200 961	(3A, 3B, 4C)	+
McL	24	2 300	2 128	2 113	78 586 201	(2A, 5B, 8A)	+
He	33	5 984	5 603	5 534	47 980 801	(2B, 3B, 7D)	+
Ru	36	7 770	7 603	7 541	1 737 216 001	(2B, 3A, 7A)	+
Suz	43	13 244	12 337	11 973	11 208 637 441	(2B, 4D, 5B)	−
ON	30	4 495	4 434	4 420	9 600 323 041	(2A, 3A, 8A)	−
Co_3	42	12 341	11 616	11 434	5 901 984 001	(2B, 3C, 7A)	+
Co_2	60	35 990	33 351	32 266	1 602 478 080 001	(2C, 3A, 11A)	−
Fi_{22}	65	45 760	41 932	39 811	768 592 281 601	(2C, 3D, 7A)	+
HN	54	26 235	25 563	25 372	3 250 368 000 001	(2B, 3B, 7A)	+
Ly	53	24 804	24 543	24 506	616 252 131 000 001	(2A, 3B, 7A)	+
Th	48	18 424	18 302	18 268	1 080 308 855 808 001	(2A, 3C, 7A)	+
Co_1	101	171 700	166 067	163 969	103 944 420 163 584 001	(2B, 4F, 5C)	−
Fi_{23}	98	156 849	149 950	146 486	85 197 301 526 937 601	(2C, 3D, 8C)	−
J_4	62	39 711	39 470	39 390	1 033 042 512 453 304 321	(2B, 3A, 7A)	+

TABLE 6. Small Genera for a Number of Sporadic Simple Groups

for relatively few class structures $C = (C_1, C_2, C_3)$ did the character theoretic criteria not admit a decision whether or not the group is C-generated. But the minus signs show that they are not good enough in general. In [**CWW92**] and [**KPW87**], more details of the subgroup structure are used in order to exclude further possibilities, which are $(2, 3, 7)$- and $(2, 3, 8)$-generation for Suz, $(2, 3, 7)$-generation for ON and Fi_{23}, and $(2, 3, 9)$-, $(2, 4, 5)$-, and $(2, 3, 10)$-generation for Co_2, and to prove $(2, 3, 8)$-generation of Co_1. Thus [**CWW92**] yields that each genus in Table 6 except the one for Co_1 is in fact the strong symmetric genus.

The corresponding genus is in all except two cases smaller than $1 + |G|/12$, thus only *triples* of classes need to be considered by Lemma 3.18 (b). In the exceptional cases M_{23} and McL, the quadruple $(2, 2, 2, 3)$ also had to be checked and was excluded by Corollary 15.10.

EXAMPLE 16.1. In general, the strong symmetric genus of a group G can arise from an epimorphism $\Gamma(g_0; m_1, m_2, \ldots, m_r) \to G$ where g_0 is strictly positive (see [**Mac65**]).

Consider the abelian group $G = 2^2 \times 6 \times 12$ of order 288. There is a surface kernel epimorphism $\Gamma(1; 2, 2, 2) \to G$ that maps the two hyperbolic generators to elements of orders 6 and 12, respectively, and maps the three elliptic generators to the three involutions in a subgroup of type 2^2. The kernel corresponds to a surface of genus

$$1 + \frac{|G|}{2} \sum_{i=1}^{3} (1 - 1/2) = 1 + \frac{|G|}{2} \cdot \frac{3}{2} = 217.$$

Suppose that G is a factor of a group $\Gamma(0; m_1, m_2, \ldots, m_r)$ such that the kernel corresponds to a surface of equal or smaller genus. This genus is

$$1 - |G| + \frac{|G|}{2} \sum_{i=1}^{r} (1 - 1/m_i) = 1 + \frac{|G|}{2} \cdot \left(r - 2 - \sum_{i=1}^{r} \frac{1}{m_i} \right),$$

so we have $(r - 2 - \sum_{i=1}^{r} (1/m_i)) \leq 3/2$ or equivalently $\sum_{i=1}^{r} (1/m_i) \geq r - 7/2$.

But $r \leq 4$ is impossible by Lemma 3.18, and $r \geq 7$ is impossible because all m_i are at least 2. For the remaining cases, note that at least three of the m_i must be divisible by 3 and at least two by 4 because of Theorem 9.1.

If $r = 6$ then the bound $r - 7/2 = 5/2$ cannot be attained under these conditions because two summands $1/4$ would force the other four summands to be $1/2$. If $r = 5$ then all m_i must be even, and with three periods that are divisible by 6, the bound $r - 7/2 = 3/2$ leaves no space for two multiples of 4.

Thus the strong symmetric genus of G is 217.   ⋄

17. Admissible Signatures

In this section, we discuss nongeneration in terms of signatures.

For given integers $g \geq 2$ and m, there exist only finitely many signatures $(g_0; m_1, m_2, \ldots, m_r)$ with $g - 1 = m(g_0 - 1) + (m/2) \sum_{i=1}^{r} (1 - 1/m_i)$ such that all periods m_i divide m. By Theorem 3.12, for each such signature there is a Fuchsian group $\Gamma = \Gamma(g_0; m_1, m_2, \ldots, m_r)$. But in many cases there is no surface kernel epimorphism from Γ onto a group of order m.

For example, choose $g = 2$ and $m = 15$. The signature $(0; 3, 3, 5)$ solves the above equation, but the group $\Gamma(0; 3, 3, 5)$ has no factor group of order 15 because every group of order 15 is abelian (in fact cyclic), and the signature does not satisfy the conditions of Theorem 9.1.

So there are combinatorially possible signatures that need not be considered for given values of g or m. In this section, we develop conditions to discard unnecessary signatures.

17.1. General Admissibility. Our definition of admissibility of a signature is pragmatic.

DEFINITION 17.1. Let m be a positive integer. We say that the signature $(g_0; m_1, m_2, \ldots, m_r)$ is *admissible for* m if none of the criteria stated below in Lemmas 17.2, 17.4, 17.6, 17.10, 17.11, Corollary 17.12, Lemmas 17.14, 17.15, and Corollary 17.16 allows us to conclude that there is no surface kernel epimorphism from $\Gamma(g_0; m_1, m_2, \ldots, m_r)$ onto a group of order m.

Note that in Lemma 17.6 and Corollary 17.16, all these criteria are applied recursively to groups of smaller order, so the well-definedness of admissibility follows by induction on m.

The following lemma restates those of the facts that we have met already and that admit a formulation in terms of signatures.

LEMMA 17.2. *Let m be a positive integer, $\Gamma = \Gamma(g_0; m_1, m_2, \ldots, m_r)$ be a Fuchsian group, and $g = 1 + m(g_0 - 1) + (m/2) \sum_{i=1}^{r}(1 - 1/m_i)$.*

(a) *Following Theorem 9.1, it can be decided whether Γ has an* abelian *surface kernel factor of order m; if yes then the signature of Γ is clearly admissible for m.*

(b) *If m is among the periods m_i then the signature of Γ is admissible for m if and only if Γ satisfies the conditions of Corollary 9.4.*

Now assume that the signature of Γ is admissible for m. Then the following statements hold.

(c) *All prime divisors of m and all periods m_i are orders of automorphisms of compact Riemann surfaces of genus g. (Note that the set of these orders can be computed by Corollary 9.4.)*

(d) *If $g_0 = 0$ and $m \equiv 2 \pmod 4$ then some of the periods m_i are even by Lemma 15.5 (b).*

(e) *If $g_0 = 0$ and $m/2$ is among the periods m_i then at least two of the other periods must be even by Lemma 15.5 (c).*

(f) *If all groups of order m are abelian then $r \neq 1$. This condition is satisfied for example if m is of the form p, p^2, pq or p^2q where p and q are primes such that $p < q$ and p does not divide $q - 1$.*

EXAMPLE 17.3. One can improve part (c) of Lemma 17.2 a little bit. Specifically, the periods m_i of an admissible signature in genus g must occur as *periods* of an admissible signature of a cyclic group in genus g. This is because for the image σ of an elliptic generator of order m_i under a surface kernel epimorphism, the full preimage $\Phi^{-1}(\langle \sigma \rangle)$ is a Fuchsian group with a period m_i and a cyclic surface kernel factor of order m_i.

With this improved criterion, the signature $(0; 2, 5, 15)$ is shown to be non-admissible in genus 8 (i.e., for the order 60) because every (cyclic) group of order 15 in genus 8 belongs to the signature $(0; 3, 3, 5, 5)$.   ◇

LEMMA 17.4. *Let p be a prime, and n, g_0 positive integers. Then $\Gamma(g_0; pn)$ has no surface kernel factor of order $p^2 n$.*

Proof. Suppose that G is a surface kernel factor of $\Gamma = \Gamma(g_0; pn)$ such that $|G| = p^2 n$. The unique elliptic generator of Γ is a product of commutators in Γ, so its image in G generates a subgroup U of G with $U \leq G'$. If G is not perfect then we have in fact $U = G'$ since U has prime index in G; but then U is normal in G, it has a characteristic subgroup of index p, hence G has an (abelian) factor group of order p^2, contrary to the fact that $U \leq G'$.

So we have to show that G cannot be perfect. For that, we consider the permutation action of G on the p right cosets of U. The factor group $\overline{G}$ modulo the kernel of this action has a Sylow p-subgroup P of order p, and it contains a cyclic subgroup $\overline{U}$ of index p. So $\overline{G} = \overline{U} \cdot P$ is a product of two cyclic groups, and this implies that $\overline{G}$ is supersolvable by a result of Itŏ, see [**Hup83**, Satz VI.10.1].   □

EXAMPLE 17.5. Let $p = n = 3$, then $\Gamma_1 = \Gamma(1; 9)$ has no surface kernel factor of order 27 by Lemma 17.4, this means that Γ_1 has no torsion-free normal subgroup of orbit genus 13. As a consequence, $\Gamma_2 = \Gamma(0; 2, 2, 2, 18)$ also has no torsion-free normal subgroup of orbit genus 13, i.e., of index 54. This is because the unique subgroup of index 2 in such a factor would not contain any of the elliptic generators of Γ_2, and so its preimage in Γ_2 would be isomorphic to Γ_1. Similarly, it follows that $\Gamma(0; 2, 4, 36)$ has no surface kernel factor of order 108.   ◇

The following observation can be used to deduce the nonadmissibility of a signature for groups of a given order from the nonadmissiblility of related signatures for groups of smaller orders. This means that the definition of admissibility becomes recursive.

LEMMA 17.6. *Let Γ be a Fuchsian group, and m a positive integer. If no perfect group of order m exists then the signature of Γ can be admissible for m only if there are a prime p dividing $\gcd(m, [\Gamma : \Gamma'])$ and a normal subgroup of index p in Γ whose signature is admissible for m/p.*

Note that this lemma can be applied only for Fuchsian groups Γ of orbit genus 0, and in this case we can compute from the signature of Γ the signatures of the normal subgroups in question by Lemma 3.6.

REMARK 17.7. A classification of perfect groups of small order has been given in [**HP89**], the groups are available as a database in GAP. From this, we read off that exactly the following numbers below $5\,000$ occur as orders of perfect groups.

$$1, 60, 120, 168, 336, 360, 504, 660, 720, 960,$$
$$1\,080, 1\,092, 1\,320, 1\,344, 1\,920,$$
$$2\,160, 2\,184, 2\,448, 2\,520, 2\,688,$$
$$3\,000, 3\,420, 3\,600, 3\,840,$$
$$4\,080, 4\,860, 4\,896.$$

EXAMPLE 17.8. Clearly each signature $(0; m_1, m_2, \ldots, m_r)$ with pairwise coprime periods m_i is not admissible for m if no perfect group of order m exists. This excludes for example the signature $(0; 2, 3, 7)$ for

$$m \in \{84, 252, 420, 588, 672, 756, 840, 924,$$
$$1\,008, 1\,176, 1\,260, 1\,428, 1\,512, 1\,596, 1\,680\};$$

in other words, the Hurwitz bound $84(g-1)$ cannot be attained for the genera

$$g \in \{2, 4, 6, 8, 9, 10, 11, 12, 13, 15, 16, 18, 19, 20, 21\}.$$

◇

EXAMPLE 17.9. The signatures $(0; 2, 3, 10)$ and $(0; 2, 5, 6)$ are not admissible for 180 (corresponding to genus 7 and 13, respectively), because there is no perfect group of order 180 by the result cited in Remark 17.7, and no $(0; 3, 3, 5)$-group of order 90 by Lemma 17.2 (d).   ◇

Table 7 shows some nonadmissible signatures for the given values of m.

m	g	$(g_0; m_1, m_2, \ldots, m_r)$	Excluded by	
12	2	$(0; 2, 4, 12)$	17.2 (b)	$(m_3 = m)$
84	2	$(0; 2, 3, 7)$	17.2 (c)	$(m_3 = 7)$
36	2	$(0; 2, 3, 9)$	17.2 (c)	$(m_3 = 9)$
6	3	$(0; 3, 3, 3, 3)$	17.2 (d)	
$4n$	$2n - 4$	$(0; n, n, 2n)$	17.2 (e)	for odd n
$4n$	$2n - 3$	$(0; n, 2n, 2n)$	17.2 (e)	for odd n
15	2	$(1; 3)$	17.2 (f)	
60	8	$(0; 2, 5, 15)$	17.3	
$4n$	$2n$	$(1; 2n)$	17.4	
180	7	$(0; 2, 3, 10)$	17.9	

TABLE 7. Nonadmissible Signatures

17.2. Admissibility for Nonsolvable Groups.
The following criteria are used to show that no surface kernel epimorphism from a Fuchsian group Γ onto a *solvable* group of given order m exists. This helps at least to reduce

the number of candidates, i.e., of isomorphism types of groups of order m, that must be considered as possible epimorphic images of Γ.

Moreover, if *all* groups of order m are solvable then the signature of Γ is proved to be nonadmissible for m in such a case. A necessary and sufficient condition for this is that m is not divisible by the order of a nonabelian simple group.

The probably most obvious criterion in this context is the following.

LEMMA 17.10. *All factor groups of a perfect group are perfect, too. Hence a signature $(0; m_1, m_2, \ldots, m_r)$ with pairwise coprime periods m_i is not admissible for solvable groups of arbitrary order.*

So let us look at groups with nontrivial solvable factors for the rest of this section. We start with a general observation and its application to Fuchsian groups.

LEMMA 17.11. *Let G be a finite solvable group and p a prime such that the Sylow p-subgroup of G is cyclic and p does not divide $[G{:}G']$. Then G has a nonabelian factor group of order ps where $s \neq 1$ is an integer that divides $p-1$ and divides the maximal order of cyclic factor groups of G.*

Proof. We choose a normal series $1 \leq N < M \leq G' < G$ such that $[M{:}N] = p$ and $[G'{:}M]$ is coprime to p. Without loss of generality, we may assume that $N = 1$. The group $C = C_G(M)$ is normal in G, and $G/C = N_G(M)/C_G(M)$ is isomorphic to a subgroup of $\mathrm{Aut}(M) \cong C_{p-1}$, thus $G' \leq C$. By the Zassenhaus Theorem [**Hup83**, Satz I.18.1], M has a complement K in G. Note that $C \neq G$, because otherwise M would be central in G, so K would be a normal subgroup of index p in G, contrary to our assumption on G'. Set $s = [G{:}C]$, then s divides $p-1$ and the maximal order of cyclic factor groups of G. Let X be a complement of M in C. Then X is characteristic in C, thus it is normal in G, and G/X is the required nonabelian factor group of order ps. □

COROLLARY 17.12. *Let m be a positive integer, and $\Gamma = \Gamma(0; m_1, m_2, \ldots, m_r)$ a Fuchsian group. Let p be a prime that does not divide $[\Gamma{:}\Gamma']$. Assume that p divides m exactly once or that one of the m_i is divisible by the p-part of m. Let i denote the maximal order of cyclic factor groups of Γ, and set $s = \gcd(p-1, m, i)$. Then Γ has no surface kernel epimorphism onto a solvable group of order m if one of the following conditions is satisfied:*

   (a) *$s = 1$,*
   (b) *$r - 2$ periods are coprime to ps,*
   (c) *$s = 2$, all periods are coprime to p, and $r - 3$ periods are coprime to ps.*

Proof. Suppose Γ has a solvable epimorphic image of order m, with torsion-free kernel. This image satisfies the assumptions of Lemma 17.11, so $s \neq 1$, and Γ has a nonabelian factor group G of order ps.

Consider an epimorphism onto G. Supposed that $r-2$ periods are coprime to ps, the images of the corresponding generators must be trivial, so the images of the remaining two generators are inverses of each other, hence G is cyclic. Contradiction.

Suppose that $s = 2$, and all periods are coprime to p, and $r-3$ periods are coprime to ps. Then any epimorphism onto G maps $r-3$ generators of Γ to the identity, and the other three generators to involutions. But the product of two involutions in G lies in the normal subgroup of order p. Contradiction. $\square$

REMARK 17.13. If $m \equiv 2 \pmod 4$, the case that $p = 2$ does not divide $[\Gamma{:}\Gamma']$, or equivalently that all periods of Γ are odd, was considered already in part (b) of Lemma 15.5. Note that this lemma did not assume that the group of order m is solvable. (But a group of order congruent to 2 modulo 4 is solvable by the Feit–Thompson Theorem.)

With the same idea as above, we can handle the following slightly different situations.

LEMMA 17.14. *If G is a solvable group with derived subgroup of prime index p then the p-part of $|G|$ cannot be p^2.*

Proof. Suppose the p-part of $|G|$ is p^2, there is a normal series $1 \leq N < M \leq G' < G$ with $[M{:}N] = p$. Without loss of generality assume $N = 1$, and observe that $G/C_G(M)$ is a cyclic group of order dividing $p-1$. Thus M is central in G, and it has a complement K in G' that is normal in G. The index of K in G is p^2, hence $G' \leq K$. Contradiction. $\square$

LEMMA 17.15. *Let m be a positive integer. Let Γ be a group, and n the index of its derived subgroup. Assume that $\gcd(m, n)$ is a power of the prime p, and that p does not divide $q - 1$ for any prime power q dividing m. If Γ has a solvable factor group of order m then m is a power of p.*

Proof. Suppose that G is a solvable factor group of Γ, of order m, and that m is not a power of p. Then there is a normal series $1 \leq N < M \leq G' < G$ such that G/M is a p-group and $[M{:}N] = q$ is a prime power coprime to p. Without loss of generality, we may assume that $N = 1$ and that G/M acts irreducibly on M. This action cannot be trivial because then M would be central in G, which is impossible as we saw in the proof of Lemma 17.11. So the action is especially fixed point free, and as p divides the length of each orbit on $M^{\times}$, it must divide $q - 1$, contrary to our assumptions. $\square$

Similarly to the statement of Lemma 17.6, the following corollary allows us to deduce the nonadmissibility of a signature for *solvable* groups of a given order from the nonadmissiblility of related signatures for *solvable* groups of smaller orders.

COROLLARY 17.16. *Let* Γ *be a Fuchsian group, and* m *a positive integer. If* Γ *has a solvable epimorphic image of order* m, *with torsion-free kernel, then there are a prime* p *dividing* $\gcd(m, [\Gamma:\Gamma'])$ *and a normal subgroup of index* p *in* Γ *whose signature is admissible for a solvable epimorphic image of order* m/p, *with torsion-free kernel.*

We illustrate the corollary with an example.

EXAMPLE 17.17. Let m be a positive integer such that $m/12 = p$ is a prime that is congruent to 0 or 2 modulo 3. We show that the group $\Gamma = \Gamma(0; 2, 4, 12)$ has no solvable surface kernel factor group of order m.

Suppose Γ has a solvable surface kernel factor group G of order m. The kernel has orbit genus g, with

$$g - 1 = -m + (m/2)\left(1 - 1/2 + 1 - 1/4 + 1 - 1/12\right) = m/12 = p.$$

The commutator factor group of G is a subgroup of 2×4, so there is a subgroup H of index 2 in G, which is a surface kernel factor of a group with one of the signatures $(0; 2, 12, 12)$, $(0; 4, 4, 6)$, $(0; 2, 2, 2, 6)$. But the first two signatures cannot occur in our situation because $|H|$ is not divisible by 4.

Supposing that H is a solvable surface kernel factor of a group with signature $(0; 2, 2, 2, 6)$, then H has a subgroup K of index 2, which is a surface kernel factor of a group with one of the signatures $(0; 2, 2, 6, 6)$, $(0; 2, 2, 2, 2, 3)$, $(1; 3)$. The first two are impossible because $|K| = 3p$ is odd, and the third can be excluded because K is abelian.   ◇

Of course we can generalize Lemma 17.15 to the situation that $\gcd(m, n) = pq$ where Corollary 17.16 excludes the prime factors of q. Here is an example.

EXAMPLE 17.18. We show that no factor group of $\Gamma = \Gamma(0; 2, 5, 10)$ has order 40.

Suppose G were such a group, then the size of the commutator factor group of G would be a divisor of 10. The preimage in Γ of a hypothetical subgroup of index 2 in G would have signature $(0; 5, 5, 5)$, and this signature is not admissible for a group of order 20 by Lemma 17.15.

So the commutator factor group of G has order 5, and Lemma 17.15 can be applied to G.   ◇

Table 8 shows some signatures that are admissible only for nonsolvable groups of the given orders.

m	g	$(g_0; m_1, m_2, \ldots, m_r)$	Excluded by	
924	12	$(0; 2, 3, 7)$	17.10	
264	12	$(0; 3, 3, 4)$	17.12 (a)	$(p = 11, s = 1)$
280	8	$(0; 2, 4, 5)$	17.12 (b)	$(p = 7, s = 2)$
120	6	$(0; 2, 4, 6)$	17.12 (c)	$(p = 5)$
60	6	$(0; 2, 2, 2, 3)$	17.12 (c)	$(p = 5)$
63	7	$(0; 3, 3, 7)$	17.14	
20	2	$(0; 2, 5, 5)$	17.15	
60	6	$(0; 2, 4, 12)$	17.17	
40	5	$(0; 2, 5, 10)$	17.18	

TABLE 8. Signatures Nonadmissible for Solvable Groups

CHAPTER 5

Classification for Small Genus

In this chapter, we consider the problem of determining all characters that come from compact Riemann surfaces of a fixed genus $g \geq 2$. In other words, we are interested in the classification of all groups of automorphisms of compact Riemann surfaces X of genus g, up to equivalence of their actions on $\mathcal{H}^1(X)$.

Our strategy is described in Section 18. It follows partly the one used for example in [**Kur84, Kur86, KK90a, KK90b**], but takes advantage of recent progress in the classification of groups of small order (see [**BE99**]). Section 19 shows an example, the classification of all *irreducible* characters that come from Riemann surfaces, and Section 20 reports on the classification of characters that come from Riemann surfaces of genus g, for $2 \leq g \leq 48$.

18. Algorithms

In order to prove whether a *given* G-character comes from a Riemann surface, one must prove the existence or nonexistence of a surface kernel epimorphism that induces this character. But for listing all characters of G that come from Riemann surfaces of a given genus g, we do not want to check all G-characters of degree g. Instead, we distribute the possible surface kernel epimorphisms to G successively into smaller equivalence classes, according to signature and distribution of generator images to conjugacy classes of G. This has the advantage that in each step, we may have the possibility of rejecting whole equivalence classes, using the methods of Chapter 4.

In a sense, we work far enough towards the classification of surface kernel epimorphisms onto G, with kernel of orbit genus g, to obtain representatives under the equivalence relation Tr (see Definition 13.1). The characters themselves will occur only at the end of the process, as output.

18.1. A Straightforward Algorithm. The idea of the following algorithm is straightforward.

ALGORITHM 18.1. For a fixed integer $g \geq 2$, we compute all characters that come from Riemann surfaces of genus g.

0. Initialize the list with the degree g character of the trivial group, corresponding to the signature $(g, -)$.

1. Loop over the possible orders n of nontrivial automorphism groups, i.e., $2 \leq n \leq 84(g - 1)$.

2. For each such n, loop over all signatures $(g_0; m_1, m_2, \ldots, m_r)$ that are admissible for n in the sense of Chapter 4 and such that $0 \leq g_0 < g$, all m_i divide n, and $g - 1 = n(g_0 - 1) + (n/2) \sum_{i=1}^{r} (1 - 1/m_i)$.

3. For each order n and each admissible signature, loop over all groups G of order n, up to isomorphism. If the signature is admissible only for nonsolvable groups of order n then consider only nonsolvable groups.

4. For each admissible signature and each group G, loop over all class structures $C = (C_1, C_2, \ldots, C_r)$ of G where C_i consists of elements of order m_i.

5. If $\mathrm{Epi}_C(g_0, G) \neq \emptyset$ then add the character induced by the epimorphisms to the list to be returned.

Now let us try to improve the steps of this algorithm.

In Step 1, clearly one can discard the numbers n between $48(g - 1)$ and $84(g - 1)$ and most other large values of n by Lemma 3.18.

Note that only *admissible* signatures are considered in Step 2. Following Lemma 17.2, in particular we may compute the possible element orders of automorphisms first (see Corollary 9.4), and choose the periods m_i from the intersection of this set with the set of divisors of m.

Before we enter Step 4 with a group G and a signature, we may test whether the abelian invariants of G and the Fuchsian group are compatible, see Section 15.2. For abelian G, we can discard the group directly if $n > 4g + 4$, and otherwise decide the existence of a surface kernel epimorphism with Theorem 9.1.

By Lemma 3.3 (see also Remark 14.2), we may fix a total ordering of the conjugacy classes and consider in Step 4 only ordered r-tuples of classes. Additionally, we can loop over class structures modulo the action of $\mathrm{Aut}(G)$ since in fact we are interested in orbits of characters under the action of this group.

Step 3 is the crucial one, because it assumes a classification of all groups of order n, up to isomorphism. This is in principle a combinatorial problem, which could be solved for example by computing equivalence classes of the finitely many multiplication tables of size n. But this approach is limited to small groups, hence small genus g. Using sophisticated methods, a classification of groups of orders up to $1\,000$ (except the orders 512 and 768) has been given in [**BE99**]. The groups are available as a GAP library, so this covers nearly all possible groups of automorphisms of Riemann surfaces of genus up to $\lfloor 1 + 1\,000/84 \rfloor = 12$.

If we treat the case of $(2, 3, 7)$-generation separately, Lemma 3.18 yields that in all other cases, the order of automorphism groups is bounded by $1\,000$ if the genus is at most $\lfloor 1 + 1\,000/48 \rfloor = 21$. Note that the Hurwitz groups of order less than one million are known by [**Con87**].

It should be mentioned that further shortcuts are possible. For example, the GAP database of 2-groups of order up to 256, which has been computed by E. A. O´Brien (see [**NO89, O´Br91**]), allows us to loop over the groups of prescribed rank. The search for surface kernel factors of $\Gamma(0; 2, 4, 8)$ of order 256 can thus be restricted to those 540 groups of order 256 that have rank 2, instead of all 56 092 groups of order 256. In this example, 119 of these groups have compatible abelian invariants, and exactly 4 of them do really occur as surface kernel factors.

18.2. A Self-Contained Algorithm. We are also interested in the classification for genera larger than 21. For that and for the cases of groups of order 512 and 768, we need an algorithm that is independent of the results given in [**BE99**]. Therefore we modify Algorithm 18.1. The idea is to proceed inductively, that is, to assume the classification of automorphism groups of genera from 2 to $g - 1$. So our approach is based on the following observation.

LEMMA 18.2. *Let g be an integer, $g \geq 2$, and $(g_0; m_1, m_2, \ldots, m_r)$ a signature. If G is a finite group such that there is a surface kernel epimorphism Φ from $\Gamma(g_0; m_1, m_2, \ldots, m_r)$ onto G, with kernel of orbit genus g, then one of the following cases occurs.*

(1) *G is perfect.*
(2) *G has a normal subgroup N of prime index p, and either*
(2a) *G is isomorphic to a subgroup $\widetilde{G}$ of $\mathrm{Aut}(N)$, with $\mathrm{Inn}(N) < \widetilde{G}$ of index p, or*
(2b) *G is isomorphic to the direct product of N and a cyclic group of order p, or*
(2c) *G has an elementary abelian normal subgroup A of order q^t, for a prime q, with A central in N and G/A a surface kernel image of a group $\Gamma(g_0; \widetilde{m}_1, \widetilde{m}_2, \ldots, \widetilde{m}_r)$, where $\widetilde{m}_i \in \{m_i/q, m_i\}$, and for which the kernel has orbit genus strictly smaller than g.*

Proof. Assume that G is not perfect. Then G has a normal subgroup N of prime index p. If the center $Z(N)$ of N is trivial then either $C_G(N) \subseteq N$ or $C_G(N) \not\subseteq N$. The former means that $C_G(N) = C_N(N) = Z(N)$, which is trivial by our assumption, thus the homomorphism that maps $\sigma \in G$ to the conjugation homomorphism $(n \mapsto n^\sigma) \in \mathrm{Aut}(N)$ is injective, hence case (2a) applies. The latter means that $C_G(N) = \langle C_N(N), \sigma \rangle = \langle \sigma \rangle$ for some element $\sigma \in G \setminus N$, and since $G = \langle N, \sigma \rangle$, we have $\sigma \in Z(G)$, thus $\langle \sigma \rangle$ is a normal subgroup of G that intersects N trivially, hence case (2b) applies.

It remains to consider the case that $|Z(N)| > 1$. Let A be a nontrivial characteristic subgroup of $Z(N)$, without loss of generality we can choose A an elementary abelian q-group. Then A is normal in G, and we may construct a surface kernel epimorphism $\widetilde{\Phi}$ from a suitable group $\Gamma(g_0; \widetilde{m}_1, \widetilde{m}_2, \ldots, \widetilde{m}_r)$ onto G/A by taking $\Phi(c_i) \cdot A$ as image of the i-th elliptic generator, and choosing the images of the hyperbolic generators analogously. Note that this defines the orders $\widetilde{m}_i$. The kernel of $\widetilde{\Phi}$ has orbit genus

$$
\begin{aligned}
\widetilde{g} &= 1 + |G/A|(g_0 - 1) + \frac{|G/A|}{2} \sum_{i=1}^{r} (1 - 1/\widetilde{m}_i) \\
&\leq 1 + \frac{1}{|A|} \left(|G|(g_0 - 1) + \frac{|G|}{2} \sum_{i=1}^{r} (1 - 1/m_i) \right) \\
&= 1 + \frac{1}{|A|} (g - 1) \leq 1 + \frac{1}{2}(g - 1) = \frac{1}{2}(g + 1) < g,
\end{aligned}
$$

that is, case (2c) applies. $\qquad \square$

REMARK 18.3. It should be noted that in the case (2c) of Lemma 18.2, it may happen that the group G/A or the signature $(g_0; \widetilde{m}_1, \widetilde{m}_2, \ldots, \widetilde{m}_r)$ belongs to genus 0 or 1. But this causes no problem by the discussion in Section 3.4. That is, if the genus is 0 then G/A is uniquely determined. In genus 1, there are only finitely many cases to consider because G/A is a surface kernel factor of a group $\mathbb{Z}^2 : m$, for $m \in \{1, 2, 4, 6\}$, and thus G/A is a factor of the finite group $|G/A|^2 : m$.

We can turn this into a self-contained algorithm for given g if we are able to compute the perfect groups of order at most $84(g - 1)$, the automorphism group of a given finite group, and all downward extensions of a given finite group by an elementary abelian group, up to isomorphism. Alternatively, the case (2c) can also be settled if we can compute all *upward* extensions of a given finite group by a cyclic group of prime order, up to isomorphism.

The case of small perfect groups is no problem because of the GAP database based on [**HP89**], see also Remark 17.7. Implementations for the other tasks are available in GAP. So our algorithm may look as follows.

ALGORITHM 18.4. For a fixed integer $g \geq 2$, we take as input all characters that come from Riemann surfaces of the genera from 2 to $g - 1$, and all perfect groups up to order $84(g - 1)$, up to isomorphism. We compute all characters that come from Riemann surfaces of genus g.

0. Initialize the list with the degree g character of the trivial group, corresponding to the signature $(g, -)$.
1. Loop over the possible orders n of nontrivial automorphism groups, i.e., $2 \leq n \leq 84(g - 1)$.
2. For each such n, loop over all signatures $(g_0; m_1, m_2, \ldots, m_r)$ that are admissible for n in the sense of Chapter 4 and such that $0 \leq g_0 < g$, all m_i divide n, and $g - 1 = n(g_0 - 1) + (n/2) \sum_{i=1}^{r} (1 - 1/m_i)$.
3. For each order n and each isomorphism type of groups Γ whose signature is admissible for n, compute the candidates G of order n as follows. First take the perfect groups G of order n, up to isomorphism. Then compute, for each prime divisor p of $\gcd(n, [\Gamma{:}\Gamma'])$, all normal subgroups Γ_0 of index p in Γ and all surface kernel factors N of Γ_0, of order n/p; we know these groups N by induction on n. If the signature of Γ is admissible only for nonsolvable groups of order n then we must consider only nonsolvable groups N.

   If N is centerless then take the direct product G of N with a cyclic group of order p and all subgroups G of $\mathrm{Aut}(N)$ that contain $\mathrm{Inn}(N)$ of index p.

   If N has a nontrivial center then compute one nontrivial characteristic elementary abelian central subgroup A of N of q-power order; then compute all surface kernel images $\widetilde{G}$ of $\widetilde{\Gamma}$, of order $n/|A|$, where $\widetilde{\Gamma}$ has signature $(g_0; \widetilde{m_1}, \widetilde{m_2}, \ldots, \widetilde{m_r})$, with $\widetilde{m_i} \in \{m_i, m_i/q\}$, and such that $\widetilde{G}$ contains a subgroup isomorphic with N/A; we know these groups $\widetilde{G}$ by induction on g. Finally, take all groups G, up to isomorphism, for which the sequence $1 \to A \to G \to \widetilde{G} \to 1$ is exact.
4. For each admissible signature and each group G, loop over all class structures $C = (C_1, C_2, \ldots, C_r)$ of G where C_i consists of elements of order m_i.
5. If $\mathrm{Epi}_C(g_0, G) \neq \emptyset$ then add the character induced by the epimorphisms to the list to be returned.

Of course the Algorithms 18.1 and 18.4 can be combined in the sense that the small groups are treated with the former and the latter is used for all cases where no loop over all isomorphism types of groups of a given order is possible. But in principle, one could work only with Agorithm 18.4, that is, without the assumption that classifications of nonperfect groups are available.

EXAMPLE 18.5 (cf. Examples 3.8, 9.11). We compute the isomorphism types of automorphism groups of compact Riemann surfaces of genus 2, starting with the 20 admissible signatures shown in Table 9. The groups corresponding to the 8 signatures with $m = m_i$ for an i are immediately identified as cyclic groups, as well as the trivial group with signature $(2; -)$. No perfect groups can occur in genus 2, and we need to consider only nonabelian groups because the abelian ones have been listed already in Example 9.11.

m	$(g_0; m_1, m_2, \ldots, m_r)$	G
1	$(2; -)$	1
2	$(0; 2, 2, 2, 2, 2, 2)$	2
2	$(1; 2, 2)$	2
3	$(0; 3, 3, 3, 3)$	3
4	$(0; 2, 2, 4, 4)$	4
4	$(0; 2, 2, 2, 2, 2)$	2^2
5	$(0; 5, 5, 5)$	5
6	$(0; 3, 6, 6)$	6
6	$(0; 2, 2, 3, 3)$	$6, S_3$
8	$(0; 4, 4, 4)$	Q_8

m	$(g_0; m_1, m_2, \ldots, m_r)$	G
8	$(0; 2, 8, 8)$	8
8	$(0; 2, 2, 2, 4)$	D_8
10	$(0; 2, 5, 10)$	10
12	$(0; 3, 4, 4)$	6.2
12	$(0; 2, 6, 6)$	2×6
12	$(0; 2, 2, 2, 3)$	D_{12}
16	$(0; 2, 4, 8)$	QD_{16}
24	$(0; 3, 3, 4)$	$SL_2(3)$
24	$(0; 2, 4, 6)$	$(2 \times 6).2$
48	$(0; 2, 3, 8)$	$GL_2(3)$

TABLE 9. Automorphism Groups in Genus 2

Each group of order 4 is abelian, so nothing is to do for the signature $(0; 2, 2, 2, 2, 2)$.

For G with signature $(0; 2, 2, 3, 3)$, possible normal subgroups N belong to the signatures $(0; 2, 2, 2, 2, 2, 2)$ and $(0; 3, 3, 3, 3)$. We have $A = N$, and we must consider downward extensions of groups of order 2 and 3; this yields the two groups of order 6.

Each normal subgroup of index 2 in $\Gamma(0; 4, 4, 4)$ has signature $(0; 2, 2, 4, 4)$, so N is cyclic of order 4, we take A of order 2 in N. The factor group N/A must have signature $(0; 2, 2)$, so G/A arises as factor of $\Gamma(0; 2, 2, 2) \cong 2^2$. We compute the nonabelian extensions of this group by A, and get both Q_8 and D_8. Only the quaternion group Q_8 is in fact $(0; 4, 4, 4)$-generated.

The signature $(0; 2, 2, 2, 4)$ is treated similarly. Candidates for N are 2^2 and 4, arising from $(0; 2, 2, 2, 2, 2)$- and $(0; 2, 2, 4, 4)$-generation, respectively. Taking A of order 2, we get G/A as factor of $\Gamma(0; 2, 2, 2)$, as above. This time the dihedral group D_8 is established.

For each $(0; 3, 4, 4)$-group G of order 12, N is of index 2 and $(0; 2, 2, 3, 3)$-generated. If $N \cong S_3$, we get no group G because $\text{Aut}(S_3) \cong S_3$; if N is cyclic then we take A of order 2, G/A is (a factor of) $\Gamma(0; 2, 2, 3) \cong S_3$, and we get a unique central extension $G \cong 6.2$ of S_3.

Three groups N must be considered for $(0; 2, 6, 6)$-groups G of order 12. First there are the signatures $(0; 2, 2, 3, 3)$ and $(0; 3, 6, 6)$, which admit both groups of order 6; we choose A of order 3, find that N/A is a $(0; 2, 2)$-group in each case, and thus G/A is a factor of $\Gamma(0; 2, 2, 2) \cong 2^2$. Only the known abelian group 2×6 is a downward extension of 2^2 that is $(2, 6, 6)$-generated. The third normal subgroup N is a $(0; 2, 2, 2, 2, 2)$-group of order 4, we choose A of order 2, thus the signature of N/A is either $(0; 2, 2, 2, 2)$ or $(0; 2, 2)$. Consequently,

G/A has signature $(0; 2, 3, 6)$ or $(0; 6, 6)$ and thus is cyclic of order 6, and again the only possible downward extension G is 2×6.

If G is a $(0; 2, 2, 2, 3)$-group of order 12 then N has signature $(0; 2, 2, 3, 3)$, which admits both groups of order 6. Taking A of order 3, we get that N/A is a $(0; 2, 2)$-group, so G/A is (a factor of) $\Gamma(0; 2, 2, 2) \cong 2^2$. The only $(0; 2, 2, 2, 3)$-group obtained as a downward extension of order 12 is the dihedral group D_{12}.

A $(0; 2, 4, 8)$-group G of order 16 is necessarily of rank 2, so it has three subgroups of index 2, with signatures $(0; 2, 2, 2, 4)$, $(0; 4, 4, 4)$, and $(0; 2, 8, 8)$. This means that we need to consider only one of these subgroups, for example $N \cong Q_8$. Taking A to be the center of N, we have $N/A \cong \Gamma(0; 2, 2, 2)$ and $G/A \cong \Gamma(0; 2, 2, 4) \cong D_8$. Among the central extensions by a group of order 2, only the quasidihedral group QD_{16} is a $(0; 2, 4, 8)$-group.

For $(0; 3, 3, 4)$, we have $(0; 4, 4, 4)$ as signature for N, so $N \cong Q_8$, and we take the center of N as A. Thus N/A is a $(0; 2, 2, 2)$-group, G/A a $(0; 2, 3, 3)$-group, hence $G/A \cong A_4$, and we get the unique $(0; 3, 3, 4)$-group $SL_2(3)$ of order 24.

Suppose G is a $(0; 2, 4, 6)$-group of order 24. Then there is a normal subgroup N whose signature is one of $(0; 2, 2, 2, 3)$, $(0; 3, 4, 4)$, $(0; 2, 6, 6)$. So N is one of D_{12}, 6.2, 2×6. In each case, we can choose A of order 2, and the signature of N/A is one of $(0; 2, 2, 3)$, $(0; 6, 6)$, $(0; 2, 3, 6)$. Thus G/A has signature $(0; 2, 2, 6)$; note that $\Gamma(0; 2, 3, 6)$ does not occur as subgroup of index 2 in a group of genus 1. We have $G/A \cong D_{12}$, and the only $(2, 4, 6)$-generated central extension of order 24 is $(2 \times 6).2$.

A $(0; 2, 3, 8)$-group G has $N \cong SL_2(3)$ of index 2, A is chosen as the center of N, so N/A is a $(0; 2, 3, 3)$-group of order 12 and thus $G/A \cong S_4 \cong \Gamma(0; 2, 3, 4)$. There are two nonisomorphic central extensions of S_4 by A, exactly one is a $(0; 2, 3, 8)$-group, namely $GL_2(3)$.

In particular, we see that each automorphism group in genus 2 occurs as a subgroup of at least one of the three groups $GL_2(3)$, $(2 \times 6).2$, and 10.   ◇

Before we apply the classification algorithms in the following sections, we need to discuss what exactly can be obtained from the output.

On the one hand, from the approach taken it is clear that we get more than the isomorphism types of the automorphism groups in genus g. Note that the signature already provides a finer distribution of the groups, as we have seen in Example 18.5, where the cyclic group of order 6 is found as both a $(3, 6, 6)$- and a $(2, 2, 3, 3)$-group in genus 2.

On the other hand, the algorithms can be modified in such a way that representatives of $\mathrm{Aut}(G)$-orbits on the sets $\mathrm{Epi}_C(g_0, G)$ are computed for the class structures C in question. If Γ is a Fuchsian group corresponding to C and

g_0 then the number of these orbits equals the number of torsion-free normal subgroups N in Γ such that $\Gamma/N \cong G$. In the case that Γ is a triangle group, we can interpret this number as the number of compact Riemann surfaces of genus equal to the orbit genus of N, with G acting as a group of automorphisms and arising from the signature of Γ. As stated in Section 13, listing the $\mathrm{Aut}(G)$-orbits of characters that come from Riemann surfaces of genus g means to give some sort of intermediate information, see Example 18.6 below.

A slightly different viewpoint is taken in [**Kur86, KK90a, KK90b**], namely, there the numbers of conjugacy classes of subgroups $\widetilde{G}$ of $GL_g(\mathbb{C})$ are counted in such a way that $\widetilde{G}$ describes the action of $G \leq \mathrm{Aut}(X)$ for a compact Riemann surface X of genus g on $\mathcal{H}^1(X)$ w.r.t. a suitable $\mathbb{C}$-basis. The subtle difference between this and counting $\mathrm{Aut}(G)$-orbits of characters, that is, the difference between considering matrix groups and matrix representations, is illustrated in Example 18.7 below.

EXAMPLE 18.6 (see Example 3.7). Let $G = \langle \sigma \rangle$ be cyclic of order 5, $\Gamma = \Gamma(0; 5, 5, 5)$, and $L = (G^\times, G^\times, G^\times)$. We have $|\mathrm{Hom}_L(\Gamma, G)| = |\mathrm{Epi}_L(\Gamma, G)| = 12$. Table 10 shows the epimorphisms (each given by the images of the generators c_1, c_2, c_3 of Γ) and the corresponding characters, where $\chi \in \mathrm{Irr}(G)$ is given by $\chi(\sigma) = \zeta_5$. Each row in the table describes an orbit under the

(c_1, c_2, c_3)	(c_1, c_2, c_3)	(c_1, c_2, c_3)	(c_1, c_2, c_3)
$(\sigma, \sigma, \sigma^3)$	$(\sigma^2, \sigma^2, \sigma)$	$(\sigma^3, \sigma^3, \sigma^4)$	$(\sigma^4, \sigma^4, \sigma^2)$
$(\sigma^3, \sigma, \sigma)$	$(\sigma, \sigma^2, \sigma^2)$	$(\sigma^4, \sigma^3, \sigma^3)$	$(\sigma^2, \sigma^4, \sigma^4)$
$(\sigma, \sigma^3, \sigma)$	$(\sigma^2, \sigma, \sigma^2)$	$(\sigma^3, \sigma^4, \sigma^3)$	$(\sigma^4, \sigma^2, \sigma^4)$
$\chi + \chi^2$	$\chi + \chi^3$	$\chi^2 + \chi^4$	$\chi^3 + \chi^4$

TABLE 10. $\mathrm{Epi}_L(\Gamma, G)$ and Characters

action of $\mathrm{Aut}(G)$, which is cyclic of order 4 and permutes $G^\times$ transitively. Clearly the character induced by an epimorphism depends only on the *unordered* triple, so each of the three torsion-free normal subgroups of Γ induces the same $\mathrm{Aut}(G)$-orbit of characters. A useful notation (which will be introduced in Chapter 6) for this and other applications is given by assigning a vector $(l(\sigma), l(\sigma^2), l(\sigma^3), l(\sigma^4))$ to an epimorphism, where $l(\sigma^i)$ denotes the number of generator images σ^i (cf. also Lemma 11.5); we get one such vector for each column of Table 10, and one orbit under the induced action of $\mathrm{Aut}(G)$ on these vectors.

In particular, if $\widetilde{G} \leq GL_2(\mathbb{C})$, $\widetilde{G} \cong G$, is induced by the action on the space $\mathcal{H}^1(X)$, for a compact Riemann surface X, then $\widetilde{G}$ is unique up to $GL_2(\mathbb{C})$-conjugacy. Note that in the computation of $\mathrm{Epi}_L(\Gamma, G)$, we need to consider only candidates $(\sigma^i, \sigma^j, \sigma^k)$ with $i \leq j \leq k$, as we stated after Algorithm 18.1. Using the transitivity of $\mathrm{Aut}(G)$, we may further restrict the search to those triples with $i = 1$, and consider a triple only if it is the smallest in its $\text{Aut}(G)$-orbit, w.r.t. the action on unordered triples. This means for example that only the triple $(\sigma, \sigma, \sigma^3)$ from the 12 triples in Table 10 must be checked.   ◇

EXAMPLE 18.7. Let $\Gamma = \Gamma(0; 3, 9, 9)$ and $G = 3_-^{1+2}$, the unique nonabelian group of order 27 and exponent 9. The character table of G is shown in Table 11, each conjugacy class is denoted by element order and a distinguishing letter, and $\zeta = \zeta_3$. Let $L_1 = 3C \cup 3D$, L_2 be the set of all elements of order 9

σ^G	1A	3A	3B	3C	3D	9A	9B	9C	9D	9E	9F
$(\sigma^3)^G$	1A	1A	1A	1A	1A	3A	3A	3A	3B	3B	3B
1_G	1	1	1	1	1	1	1	1	1	1	1
χ_2	1	1	1	ζ	ζ^2	1	ζ	ζ^2	1	ζ	ζ^2
χ_3	1	1	1	ζ^2	ζ	1	ζ^2	ζ	1	ζ^2	ζ
χ_4	1	1	1	1	1	ζ	ζ	ζ	ζ^2	ζ^2	ζ^2
χ_5	1	1	1	ζ	ζ^2	ζ	ζ^2	1	ζ^2	1	ζ
χ_6	1	1	1	ζ^2	ζ	ζ	1	ζ^2	ζ^2	ζ	1
χ_7	1	1	1	1	1	ζ^2	ζ^2	ζ^2	ζ	ζ	ζ
χ_8	1	1	1	ζ	ζ^2	ζ^2	1	ζ	ζ	ζ^2	1
χ_9	1	1	1	ζ^2	ζ	ζ^2	ζ	1	ζ	1	ζ^2
χ_{10}	3	3ζ	$3\zeta^2$	0	0	0	0	0	0	0	0
χ_{11}	3	$3\zeta^2$	3ζ	0	0	0	0	0	0	0	0

TABLE 11. Character Table of 3_-^{1+2}

in G, and consider $L = (L_1, L_2, L_2)$. We have $|\text{Hom}_L(\Gamma, G)| = |\text{Epi}_L(\Gamma, G)| = 108$; note that the third powers of elements in L_2 lie in $3A \cup 3B$. The group $\text{Aut}(G)$ has order 54, so there are two orbits on $\text{Epi}_L(\Gamma, G)$.

The elements of $\text{Epi}_L(\Gamma, G)$ induce six characters, which are of the form $\chi + \chi_{10} + \chi_{11}$, where χ is one of the six linear characters of G not containing $3C \cup 3D$ in the kernel. $\text{Aut}(G)$ induces the permutation action

$$\langle (9A, 9B, 9C)(9D, 9F, 9E), (3A, 3B)(9A, 9D)(9B, 9E, (9C, 9F) \rangle \cong S_3,$$

it has two orbits on these characters, one containing the complex conjugates of the characters in the other. In particular, complex conjugation is *not* induced by a group automorphism.

But if $\widetilde{G} = \langle M_1, M_2, M_3 \rangle \leq GL_7(\mathbb{C})$ with $\widetilde{G} \cong G$ and such that the map $\Phi_1 \colon \Gamma \to \widetilde{G}$ given by $c_i \mapsto M_i$ is a surface kernel epimorphism then also $c_1 \mapsto M_1^{-1}$, $c_2 \mapsto M_2^{-1}$, $c_3 \mapsto M_2 M_3^{-1} M_2^{-1}$ defines a surface kernel epimorphism $\Phi_2 \colon \Gamma \to \widetilde{G}$. The characters induced by Φ_1 and Φ_2 are mutually complex conjugate, so we see that the characters in the two $\text{Aut}(G)$-orbits do in fact belong to the same matrix group $\widetilde{G}$. In other words, the subgroup of $GL_7(\mathbb{C})$ isomorphic to G and corresponding to the action of an automorphism group on

$\mathcal{H}^1(X)$, for a suitable compact Riemann surface X, is unique up to $GL_7(\mathbb{C})$-conjugacy.

By the calculations in Section 20, we find that $g = 7$ is the smallest genus for which it may happen that the $\mathrm{Aut}(G)$-orbits on characters coming from Riemann surfaces are not in bijection with $GL_g(\mathbb{C})$-classes of matrix groups isomorphic with G, and belonging to these characters. So this phenomenon does not occur (and is not mentioned) in the classifications for $2 \leq g \leq 5$ in [**Kur84, Kur86, KK90b, KK90a**].   ◇

19. Irreducible Characters Coming from Riemann Surfaces

If an irreducible character of the group G comes from a Riemann surface of genus $g \geq 2$ then $g^2 < |G|$ because $|G|$ is the sum of degree squares of all irreducible characters of G. By $|G| \leq 84(g-1)$, this implies $g^2 < 84(g-1)$ and thus $2 \leq g \leq 82$. So only finitely many groups have irreducible characters that come from Riemann surfaces. In this section, we compute these characters.

In our situation, $4(g - 1) = g^2 - (g - 2)^2 \leq g^2 < |G|$, so Lemma 3.18 yields that we need to consider only signatures $(0; m_1, m_2, \ldots, m_r)$ with $3 \leq r \leq 4$. For fixed genus g, all possible group orders are multiples of g, more exactly, they are in the set $\{kg \mid g + 1 \leq k \leq 84(g - 1)/g\}$.

As sketched in Lemma 3.18, we can improve the a priori bounds on g and $|G|$ by the observation that either G arises as a factor of $\Gamma(0; 2, 3, 7)$ or $|G| \leq 48(g - 1)$. In the former case, $|G| = 84(g - 1)$, hence g divides 84 and thus is in the set $\{2, 3, 4, 6, 7, 12, 14, 21, 28, 42\}$. In the latter case, $g^2 < |G|$ forces $g \leq 46$.

We need not consider signatures $(0; m_1, m_2, \ldots, m_r)$ where an m_i is equal to the group order because all irreducible characters of the group would be linear. If $g > 2$ then those signatures with an m_i equal to half the group order can also be excluded because the group has a cyclic normal subgroup of index 2 in such a case, which implies that its irreducible degrees are bounded by 2.

Exactly 165 signatures are combinatorially possible for irreducible actions, they are determined using the facts that only triples and quadruples can occur, and that the corresponding group order is divisible by the genus. Seventy-five of them are excluded by the criteria of Section 17, and 16 more signatures cannot occur because of forbidden periods m and $m/2$ for groups of order m. So 74 remain to be considered, 30 of them only for nonsolvable groups.

These signatures are listed in Table 12, together with the information whether only nonsolvable groups have to be considered (because of the criteria from

g	$|G|$	Periods	Solv.?	G
2	6	$(2,2,3,3)$	+	D_6
	8	$(4,4,4)$	+	Q_8
		$(2,2,2,4)$	+	D_8
	12	$(3,4,4)$	+	6.2
		$(2,6,6)$	+	
		$(2,2,2,3)$	+	D_{12}
	16	$(2,4,8)$	+	QD_{16}
	24	$(3,3,4)$	+	$SL_2(3)$
		$(2,4,6)$	+	$(2\times 6).2$
	48	$(2,3,8)$	+	$GL_2(3)$
3	12	$(2,2,3,3)$	+	A_4
	21	$(3,3,7)$	+	$7:3$
	24	$(3,4,4)$	+	S_4
		$(3,3,6)$	+	
		$(2,6,6)$	+	$A_4\times 2$
		$(2,2,2,3)$	+	S_4
	48	$(3,3,4)$	+	$4^2:3$
		$(2,4,6)$	+	$2\times S_4$
		$(2,3,12)$	+	
	96	$(2,3,8)$	+	$4^2:S_3$
	168	$(2,3,7)$	−	$L_3(2)$
4	20	$(4,4,5)$	+	$5:4$
		$(2,2,2,5)$	+	
	24	$(3,4,6)$	+	
		$(2,2,2,4)$	+	
	36	$(3,4,4)$	+	$3^2:4$
		$(3,3,6)$	+	
		$(2,6,6)$	+	$S_3\times S_3$
		$(2,2,2,3)$	+	
	40	$(2,4,10)$	+	
	60	$(2,5,5)$	−	A_5
		$(2,3,15)$	−	
	72	$(2,4,6)$	+	$3^2:D_8$
		$(2,3,12)$	+	
	108	$(2,3,9)$	+	
	120	$(2,4,5)$	−	S_5
5	60	$(3,3,5)$	−	A_5

g	$|G|$	Periods	Solv.?	G
		$(2,5,6)$	−	
	80	$(2,5,5)$	+	$2^4:5$
	120	$(2,3,10)$	−	$A_5\times 2$
	160	$(2,4,5)$	+	$2^4:D_{10}$
6	48	$(3,3,8)$	+	
		$(2,6,8)$	+	
	60	$(3,4,4)$	−	
		$(3,3,6)$	−	
		$(2,6,6)$	−	
		$(2,4,12)$	−	
		$(2,2,2,3)$	−	
	72	$(2,4,9)$	+	$(2^2\times 3).S_3$
	96	$(2,3,16)$	+	
	120	$(3,3,4)$	−	
		$(2,4,6)$	−	S_5
		$(2,3,12)$	−	
	150	$(2,3,10)$	+	
	180	$(2,3,9)$	−	
	240	$(2,3,8)$	−	
7	56	$(2,7,7)$	+	$2^3:7$
	112	$(2,4,7)$	+	
	504	$(2,3,7)$	−	$L_2(8)$
8	72	$(2,4,18)$	+	
	120	$(2,3,20)$	−	
	168	$(3,3,4)$	−	$L_3(2)$
		$(2,4,6)$	−	
		$(2,3,12)$	+	
	336	$(2,3,8)$	−	$L_3(2).2$
9	144	$(2,3,18)$	+	
10	120	$(2,4,10)$	−	
	180	$(2,5,5)$	−	
		$(2,3,15)$	−	
	360	$(2,4,5)$	−	A_6
12	168	$(2,3,28)$	−	
14	1092	$(2,3,7)$	−	$L_2(13)$
15	240	$(2,3,20)$	−	
16	720	$(2,3,8)$	−	

TABLE 12. Irreducible Actions

Section 17.2), and with all groups that actually have an irreducible character that comes from a Riemann surface. The representations themselves are listed in Appendix B.

In the case that the group in question is *nonsolvable*, the identification of isomorphism type and characters is also easy without using a classification of groups. We state the proofs below to show how the criteria developed for solvable groups in Section 17.2 can also be applied to nonsolvable groups.

For solvable groups also, many signatures admit an easy identification of the groups, for example the possible automorphism groups of orders 80 and 160 in genus 5 must have types $2^4 \colon 5$ and $2^4 \colon D_{10}$, respectively, by the module criterion 17.15. Analogously, a group of order 56 in genus 7 must have type $2^3 \colon 7$. No group of order 112 is possible in genus 7 because $GL_3(2)$ has no subgroup of order 14. Other arguments to identify groups or discard signatures need the assumption that the group has an irreducible character of degree g; for example, the only surface kernel factor of order 12 of $\Gamma(0; 2, 6, 6)$ is abelian by Example 9.9, and thus has no irreducible character of degree 2.

Here are the arguments for the nonsolvable cases.

$|G| = 60$:   We have $G \cong A_5$. The period triples $(2, 5, 5)$ and $(3, 3, 5)$ lead to unique irreducible characters of degrees 4 and 5, respectively, the triples $(2, 3, 15)$ and $(2, 5, 6)$ are impossible because A_5 contains no elements of orders 15 and 6, and the triples for $g = 6$ are impossible because A_5 has no irreducible character of degree 6.

$|G| = 120$:   We have G isomorphic to one of $2.A_5$, S_5, $A_5 \times 2$. The group $2.A_5$ contains a unique involution, so by Lemma 15.6 only the triple $(3, 3, 4)$ remains for $2.A_5$, which has been shown to be nongenerating in Example 15.15. The group $A_5 \times 2$ contains no elements of orders 4, 12, 20. Only the triple $(2, 3, 10)$ is possible for $A_5 \times 2$, and it leads to a unique irreducible character of degree 5. The group S_5 contains no elements of orders 10, 12, 20, so only the triples $(2, 4, 5)$, $(2, 4, 6)$, and $(3, 3, 4)$ remain. The first two of these lead to unique irreducible characters of degrees 4 and 6, respectively, and the last is impossible because the generator of order 4 would be the only one outside the normal subgroup A_5.

$|G| = 168$:   We have $G \cong L_3(2)$. The period triples $(2, 3, 7)$ and $(3, 3, 4)$ lead to irreducible characters of degrees 3 and 8, respectively, the triples $(2, 4, 6)$, $(2, 3, 12)$, and $(2, 3, 28)$ are impossible because $L_3(2)$ contains no elements of orders 6, 12, and 28.

$|G| = 180$:   We have $G \cong A_5 \times 3$, which has no irreducible character of degree 6 or 10.

$|G| = 240$:   There is no perfect group of order 240, so the derived subgroup of G is nonsolvable of order 120 and arises either from $(3, 3, 4)$-generation, which has been excluded above, or from $(3, 3, 10)$-generation. In the latter case, G would have an irreducible character of degree 15, whose restriction to G' remains irreducible; but this is impossible because $15^2 > 120$.

$|G| = 336$:   The period triple is $(2, 3, 8)$, and we have G isomorphic to one of $2.L_3(2)$, which is excluded by Lemma 15.6 because it contains a unique involution, $L_3(2).2$, for which we get an irreducible character, and $L_3(2) \times 2$, which contains no elements of order 8.

$|G| = 360$:   Here either $G \cong A_6$, which has an irreducible character of degree 10 that comes from a Riemann surface, or G has a composition factor A_5. In the latter case, G has a composition factor of order 3 which cannot be central because $\Gamma(0; 2, 4, 5)$ has no factor group of order 3. So the centralizer of this composition factor has index 2 in G, and it is isomorphic to $A_5 \times 3$. But this group is not a factor group of $\Gamma(0; 2, 5, 5)$.

$|G| = 504$:   We have $G \cong L_2(8)$, and $(2, 3, 7)$-generation belongs to the character χ_2 (see [CCN$^+$85, p. 6]), which is the unique rational irreducible degree 7 character of $L_2(8)$.

$|G| = 720$:   Here G is $(2, 3, 8)$-generated and has a composition factor A_5 or A_6. In the case of A_6, G is one of the groups $2.A_6$, $A_6 \times 2$, $A_6.2_1$, $A_6.2_2$, $A_6.2_3$. None of these can occur, because $2.A_6$ contains a unique involution, $A_6 \times 2$ and $A_6.2_1$ contain no elements of order 8, $A_6.2_2$ has no irreducible character of degree 16, and all involutions of $A_6.2_3$ lie in its subgroup A_6. In the case of A_5, G has a cyclic composition factor of order 3, which cannot be central, thus G' has order 360 and is $(3, 4, 4)$-generated. Furthermore, the central subgroup of order 3 in G' has a complement in G', so G' has a normal subgroup of order 120 that is $(4, 4, 4)$-generated. But this group must be one of $2.A_5$, S_5, $A_5 \times 2$, and $2.A_5$ cannot occur because otherwise its factor group A_5 would be $(2, 2, 2)$-generated, S_5 is impossible because it contains no elements of order 4 in its normal subgroup A_5, and $A_5 \times 2$ contains no elements of order 4 at all.

$|G| = 1\,092$:   We have $G \cong L_2(13)$, and $(2, 3, 7)$-generation belongs to the character χ_9.

20. Calculations for $2 \le g \le 48$

With the algorithms described in Section 18, we compute all characters that come from Riemann surfaces of the genera from 2 to 48. Only the following cases cannot be treated with Algorithm 18.1, that is, with the help of the GAP library of small groups. We sketch the construction of the corresponding groups following Algorithm 18.4.

$g = 14$: $m = 1\,092$, $(0; 2, 3, 7)$. The unique perfect group of order $1\,092$ is $L_2(13)$, which is $(2, 3, 7)$-generated, see Example 15.31.

$g = 17$: $m = 768$, $(0; 2, 3, 8)$. The derived subgroup of every $(2, 3, 8)$-group G of order m is a $(3, 3, 4)$-group of order $384 = 2^7 \cdot 3$. By the catalogue of small groups (see [**BE99**]), there is a unique such group, which is centerless, so it embeds into its automorphism group of order $6\,144 = 2^{11} \cdot 3$. This automorphism group has seven classes of subgroups of order m that contain the group of inner automorphisms, which correspond to the seven classes of involutions in the group of outer automorphisms. Four of these candidates are excluded because they are nonsplit extensions of the $(3, 3, 4)$-group, two more are not $(2, 3, 8)$-generated, and the last yields a unique $(2, 3, 8)$-group of order m.

$g = 17$: $m = 1\,344$, $(0; 2, 3, 7)$. There are exactly two perfect groups of order m, both extensions of an elementary abelian group 2^3 by a group of type $L_3(2)$. The nonsplit extension is a Hurwitz group, the split extension is not.

$g = 22$: $m = 1\,008$, $(0; 2, 3, 8)$. There is no perfect group of order m, and the unique $(3, 3, 4)$-group of order 504 is $3 \times L_3(2)$. If G is a $(2, 3, 8)$-group of order m then the factor groups by the normal subgroups 3 and $L_3(2)$ are factors of $\Gamma(0; 2, 3, 8)$, so these factor groups are S_3 and $\text{Aut}(L_3(2))$, respectively. Hence $G = (3 \times L_3(2)) : 2$ is the unique subdirect product of S_3 and $\text{Aut}(L_3(2))$, which is in fact $(2, 3, 8)$-generated.

$g = 26$: $m = 1\,200$, $(0; 2, 3, 8)$. There is no perfect group of order m, and the unique $(3, 3, 4)$-group N of order 600 is of type $5^2 : SL_2(3)$, the factor group acting transitively on the 24 elements of order 5. In particular, N is centerless. The automorphism group of N has order $2\,400$, its unique normal subgroup of index 2 is *not* $(2, 3, 8)$-generated because it has a factor group of type $2.(A_4 \times 2)$. Since obviously $N \times 2$ is also not $(2, 3, 8)$-generated, we get no group in this case.

$g = 27$: $m = 2\,184$, $(0; 2, 3, 7)$. The unique perfect group of order m is $2.L_2(13)$, which cannot be $(2, 3, 7)$-generated because it has a unique involution (see Lemma 15.6).

$g = 28$: $m = 1\,080$, $(0; 2, 4, 5)$. The unique perfect group of order m is $3.A_6$, which is not $(2, 4, 5)$-generated. Furthermore, there is no nonperfect $(2, 4, 5)$-group of order m because no $(2, 5, 5)$-group of order 540 exists.

$g = 28$: $m = 1\,296$, $(0; 2, 3, 8)$. There is no perfect group of order m, and a unique $(3, 3, 4)$-group N of order 648, with center A of order 3; in fact the structure of N is $3_+^{1+2}.SL_2(3)$. The factor group N/C is a $(3, 3, 4)$-group of order 216, in genus 10, and there is exactly one $(2, 3, 8)$-group $\widetilde{G}$ of order 432. Forming all extensions of $\widetilde{G}$ with a 1-dimensional module over the field with three elements yields a unique $(2, 3, 8)$-group of order m.

$g = 29$: $m = 1\,008$, $(0; 2, 3, 9)$. There is no perfect group of order m, and the unique $(2, 2, 2, 3)$-group N of order 336 is of type $L_3(2).2$. This group is centerless and has trivial outer automorphism group, and the direct product $3 \times N$ is clearly not $(2, 3, 9)$-generated. So we get no group G in this case.

$g = 29$: $m = 1\,344$, $(0; 2, 3, 8)$. Neither of the two perfect groups of order m is $(2, 3, 8)$-generated, and there is no $(3, 3, 4)$-group of order 672. So we get no group G in this case.

$g = 31$: $m = 1\,080$, $(0; 2, 3, 9)$. The unique perfect group of order m is $3.A_6$, which is not $(2, 3, 9)$-generated, and the unique $(2, 2, 2, 3)$-group of order 360 is $N \cong A_5 \times S_3$. This group has no element of order 3 in its outer automorphism group, and clearly $N \times 3$ is not $(2, 3, 9)$-generated. So we get no group G in this case.

$g = 31$: $m = 2\,520$, $(0; 2, 3, 7)$. The unique perfect group of order m is A_7, which is not $(2, 3, 7)$-generated. So we get no group G in this case.

$g = 33$: $m = 512$, $(0; 2, 4, 8)$. Each such group is a 2-group of rank 2, so it must have three normal subgroups of index 2, and we need to consider only one of them. Let N be a $(0; 4, 4, 4)$-group of order 256 and A a characteristic elementary abelian central subgroup of N; there are exactly nine candidates for N, and the center is of order 2 or 4. The signature of N/A must be one of $(0; 2, 2, 2)$, $(0; 2, 2, 4)$, $(0; 2, 4, 4)$, $(0; 4, 4, 4)$. The first two possibilities would mean that N/A is 2^2 or D_8, which is clearly impossible; the third signature belongs to genus 1, and there is no group in genus 1 with $\Gamma(0; 2, 4, 4)$ of index 2, so this case also cannot occur. Thus N/A must be a $(0; 4, 4, 4)$-group, hence G/A a $(0; 2, 4, 8)$-group. Looking at the $(0; 2, 4, 8)$-groups of orders 256 and 128, respectively (in genus 17 and 9 respectively), we find four groups G/A in each case. Computing downward extensions that are $(2, 4, 8)$-generated, we get eight nonisomorphic groups G.

$g = 33$: $m = 768$, $(0; 3, 3, 4)$. As we have seen above, there are nine candidates of order 256 that are $(4, 4, 4)$-generated, G/A is again a $(3, 3, 4)$-group, there is exactly one of order 384 and none of order 192. Computing downward extensions, we get exactly one group G.

$g = 33$: $m = 768$, $(0; 2, 4, 6)$. Here we have three possible signatures for normal subgroups N of index 2, namely $(0; 2, 6, 6)$, $(0; 3, 4, 4)$, and $(0; 2, 2, 2, 3)$, and there are 7, 6, and 6 groups of order 384 for these signatures, with centers of order 1, 2, and 4, respectively. In the cases with trivial center, we compute the automorphism groups, in the other cases A must be a 2-group, and G/A a $(2, 4, 6)$-group. The numbers of $(2, 4, 6)$-groups of orders 384 and 192 are 4 and 2, respectively. Computing downward extensions, we get exactly six nonisomorphic groups G; in fact each of them has commutator factor group 2^2.

$g = 33$: $m = 768$, $(0; 2, 3, 12)$. For the signature $(0; 3, 3, 6)$, there are 4 subgroups with center of order 1 or 2; for $(0; 2, 2, 2, 4)$, there are 33 subgroups with center of order dividing 8. Besides the automorphism groups in the cases with trivial center, we have to look at $(2, 3, 12)$-groups of orders 384, 192, and 96 (there is exactly one such group, of order 192), and to look at $(2, 3, 6)$-groups of orders 384 and 192. We end up with 4 groups G. (It should be noted that the $1 + 6 + 4$ groups of order 768 in genus 33 are pairwise nonisomorphic.)

$g = 33$: $m = 2\,688$, $(0; 2, 3, 7)$. None of the three perfect groups of order m is $(2, 3, 7)$-generated. So we get no group G in this case.

$g = 34$: $m = 1\,320$, $(0; 2, 4, 5)$. The unique perfect group of order m is $2.L_2(11)$, which cannot be $(2, 4, 5)$-generated because it has a unique involution (see Lemma 15.6). The unique $(2, 5, 5)$-group of order 660 is $N \cong L_2(11)$, so we must consider $\mathrm{Aut}(N)$ and $N \times 2$ as candidates for G. The former is in fact $(2, 4, 5)$-generated, the latter is not.

$g = 37$: $m = 1\,080$, $(0; 2, 3, 10)$. The unique perfect group of order m is $3.A_6$, which does not contain elements of order 10, and there is no $(3, 3, 5)$-group of order 540. So we get no group G in this case.

$g = 37$: $m = 1\,296$, $(0; 2, 3, 9)$. There is no perfect group of order m. The two $(2, 2, 2, 3)$-groups N of order 432 are both centerless, the direct product $N \times 3$ cannot occur in either case, and the automorphism group of N has order 432 or $1\,728 = 4 \cdot 432$ respectively. So we get no group G in this case.

$g = 37$: $m = 1\,440$, $(0; 2, 4, 5)$. There is no perfect group of order m, and no $(2, 5, 5)$-group of order 720. So we get no group G in this case.

$g = 37$: $m = 1\,728$, $(0; 2, 3, 8)$. There is no perfect group of order m, and a unique $(3, 3, 4)$-group N of order 864, which is centerless. Since $N \times 2$ cannot be $(2, 3, 8)$-generated, we consider the subgroups of $\mathrm{Aut}(N)$, which has order $6\,912 = 8 \cdot 864$. This group has a unique normal subgroup of order 864, with factor group isomorphic with D_8. Taking preimages of the three classes of involutions in D_8, we find that exactly one is a $(2, 3, 8)$-group G of order m, which has structure $(6 \times 6) \cdot GL_2(3)$.

$g = 41$: $m = 768$, $(0; 2, 3, 16)$. There is no perfect group of order m, and there are three $(3, 3, 8)$-groups N_1, N_2, N_3 of order 384. The centers of N_1 and N_2 are trivial, and since $N_i \times 2$ cannot be $(2, 3, 16)$-generated, we consider subgroups of $\mathrm{Aut}(N_i)$. We have $|\mathrm{Aut}(N_1)| = 3\,072 = 8 \cdot 384$ and $\mathrm{Aut}(N_1)/\mathrm{Inn}(N_1) \cong 2 \times 4$; because none of the preimages of the three involutions contains elements of order 16, we get no group G from N_1. We have $|\mathrm{Aut}(N_2)| = 6\,144 = 16 \cdot 384$ and $\mathrm{Aut}(N_2)/\mathrm{Inn}(N_2) \cong 2 \times D_8$; exactly one preimage of the representatives of involution classes in this group yields a $(2, 3, 16)$-group of order m that contains a group isomorphic with N_2. The center A of

N_3 has order 2, the factor group N_3/A is either a $(3,3,4)$-group in genus 9 or a $(3,3,8)$-group in genus 21. The former leads to $(2,3,8)$-generated groups $\tilde{G}$ of order 384, but no such groups exist. In the latter case, each $\tilde{G}$ is a $(2,3,16)$-group of order 384, and there are exactly two such groups. Computing their extensions with a trivial module over the field with two elements, we find exactly one $(2,3,16)$-group G of order m.

$g = 41$: $m = 1\,440$, $(0;2,3,9)$. There is no perfect group of order m, and the unique $(2,2,2,3)$-group N of order 480 is a central product of $2.A_5$ and Q_8. Thus the factor group $A_5 \times 2^2$ is also a $(2,2,2,3)$-group, in genus 21. Since there is no $(2,3,9)$-group of order 720, we get no group G in this case.

$g = 41$: $m = 1\,920$, $(0;2,3,8)$. None of the seven perfect groups of order m is $(2,3,8)$-generated, and there is no $(3,3,4)$-group of order 960. So we get no group G in this case.

$g = 43$: $m = 1\,008$, $(0;3,3,4)$. There is no perfect group of order m, and the unique $(4,4,4)$-group N of order 336 is of type $L_3(2) \times 2$. The group $N/Z(N) \cong L_3(2)$ occurs in genus 22, extending uniquely to the $(3,3,4)$-group $L_3(2) \times 3$. Each central extension of this group by a group of order 2 has a normal cyclic subgroup of order 6, so only $N \times 3$ would be possible; but this group is not $(3,3,4)$-generated.

$g = 43$: $m = 1\,008$, $(0;2,4,6)$. There are no $(3,4,4)$- and $(2,6,6)$-groups of order 504, and $L_2(8)$ is the unique $(2,2,2,3)$-group of this order. So only $L_2(8) \times 2$ is a candidate for G, but this group has no element of order 4.

$g = 43$: $m = 1\,008$, $(0;2,3,12)$. For the signature $(0;3,3,6)$, there are two groups with center of order 6 and 3, respectively. In either case, we may choose a central subgroup of order 3, and the factor group $\tilde{G}$ is a $(2,3,12)$-group of order 336 in genus 15. There is exactly one such group, and it leads to a unique $(2,3,12)$-group of order $1\,008$; its derived subgroup has index 6, so this group is also obtained when $(2,2,2,4)$-generation is considered. For $(0;2,2,2,4)$, there are three groups of order 336 with center of order 2 and one group with trivial center. The latter group N is of type $\mathrm{Aut}(L_3(2))$. We have $N \cong \mathrm{Aut}(N)$, and $N \times 3$ cannot occur because it has a normal subgroup of type $L_3(2) \times 3$, which is not a $(3,3,6)$-group. For the former three groups, we have to consider $(2,3,12)$- and $(2,3,6)$-groups $\tilde{G}$ of order 504. There are no groups of the latter kind (in genus 1), and exactly one group of the former (in genus 22), which admits a unique central extension of order $1\,008$. Thus there is exactly one $(2,3,12)$-group of this order.

$g = 43$: $m = 1\,512$, $(0;2,3,9)$. There is no perfect group of order m, and (see above) the unique $(2,2,2,3)$-group N of order 504 is of type $L_2(8)$.

We must consider $N \times 3$ and $\mathrm{Aut}(N)$ as candidates for G. Both groups are in fact $(2, 3, 9)$-generated.

$g = 43$: $m = 1\,680$, $(0; 2, 4, 5)$. There is neither a perfect group of order m, nor a $(2, 5, 5)$-group of order 840.

$g = 43$: $m = 2\,016$, $(0; 2, 3, 8)$. There is neither a perfect group of order m, nor (see above) a $(3, 3, 4)$-group of order $1\,008$.

$g = 45$: $m = 1\,320$, $(0; 2, 3, 10)$. The unique perfect group of order m is of type $2.L_2(11)$, which cannot be $(2, 3, 10)$-generated because it has a unique involution (see Lemma 15.6). For the unique $(3, 3, 5)$-group of order 660, which is of type $L_2(11)$, we must consider $L_2(11) \times 2$ and $\mathrm{Aut}(L_2(11))$; the latter group is $(2, 3, 10)$-generated, the former is not.

$g = 46$: $m = 1\,080$, $(0; 3, 3, 4)$. The unique perfect group of order m, of type $3.A_6$, is $(3, 3, 4)$-generated. The only $(4, 4, 4)$-group of order 360 is the alternating group A_6, and $3 \times A_6$ is also $(3, 3, 4)$-generated.

$g = 46$: $m = 1\,080$, $(0; 2, 4, 6)$. The group $3.A_6$ is not $(2, 4, 6)$-generated, and there is no group of order 540 arising from one of the signatures $(0; 2, 2, 2, 3)$, $(0; 3, 4, 4)$, $(0; 2, 6, 6)$.

$g = 46$: $m = 1\,080$, $(0; 2, 3, 12)$. The group $3.A_6$ is not $(2, 3, 12)$-generated, and there is no $(3, 3, 6)$-group of order 540. One of the three $(2, 2, 2, 4)$-groups of order 360 is of type A_6, and $3 \times A_6$ is not $(2, 3, 12)$-generated; the other two $(2, 2, 2, 4)$-groups are solvable, they need not be considered because any $(2, 3, 12)$-group of order $1\,080$ is nonsolvable by Corollary 17.12 (c).

$g = 46$: $m = 1\,800$, $(0; 2, 4, 5)$. There is neither a perfect group of order m, nor a $(2, 5, 5)$-group of order 900.

$g = 46$: $m = 2\,160$, $(0; 2, 3, 8)$. The unique perfect group of order $2\,160$, of type $6.A_6$, is not $(2, 3, 8)$-generated. Any $(3, 3, 4)$-group of order $1\,080$ has a central subgroup of order 3 (see above), and the unique $(2, 3, 8)$-group of order 720, in genus 16, is of type $PGL_2(9)$. In [$\mathbf{CCN^{+}85}$, p. 4], this group is denoted by $A_6.2_2$. Thus G is either of type $3.A_6.2_2$ or a subdirect product of $A_6.2_2$ and a (nonabelian) group of order 6. The former is $(2, 3, 8)$-generated, and in the latter case also, we get one $(2, 3, 8)$-group.

Table 13 lists for each genus g, $2 \leq g \leq 48$, the number of admissible signatures (Adm.) and of signatures for which an automorphism group in genus g exists (Act.); besides the total number, the latter column lists also the numbers of signatures with zero and nonzero orbit genus, respectively. Further, Table 13 lists the number of different isomorphism types (Iso.) of automorphism groups in genus g, the number of signature–group pairs (Grp.), the number of $\mathrm{Aut}(G)$-orbits (Orb.) on G-characters that come from Riemann surfaces, the maximal order ($|G|_{\mathrm{max}}$) of an automorphism group in genus g, a corresponding signature (Max. sign.), and in the last column the quotient $q_{max} = |G|_{max}/(g-1)$.

For example, the cyclic groups of order 2 and 6 occur twice among the 19 groups in genus 2, each for two different signatures (see Table 4 on page 35).

REMARK 20.1. Table 13 does *not* list the numbers of conjugacy classes of subgroups of $GL_g(\mathbb{C})$ that describe the action of automorphism groups in genus g on the spaces $\mathcal{H}^1(X)$ w.r.t. a $\mathbb{C}$-basis. For $g \leq 6$, however, this number coincides with the number of $\mathrm{Aut}(G)$-orbits, see Example 18.7. Comparing our results with the classifications for $2 \leq g \leq 5$ in [**Kur84, Kur86, KK90b, KK90a**], we confirm that exactly 21 conjugacy classes of subgroups of $GL_g(\mathbb{C})$ occur for $g = 2$, 55 for $g = 3$, and 116 for $g = 5$. For the case $g = 4$, [**KK90b**] lists 74 instead of 73 classes of matrix groups; this is because on the one hand, the groups $H(5,20)$ and $G(5,5 \times 2)$ of order 5 in [**KK90b**, Proposition 2.1 (d)] and hence the groups $H(5 \times 2, 10)$ and $G(5 \times 2)$ of order 10 in [**KK90b**, Proposition 2.3 (a)] are equal, and on the other hand one orbit for the group $S_3 \times 3$ of order 18, for the signature $(0; 3, 6, 6)$, is missing in [**KK90b**, Proposition 2.4 (c)].

REMARK 20.2. We see that the maximal order of automorphism groups is particularly small for $g \in \{18, 23, 24, 30, 47, 48\}$. Specifically, this maximal order is equal to $8(g+1)$ for $g \in \{23, 47\}$ and equal to $8(g+3)$ for $g \in \{18, 24, 30, 48\}$. In [**Acc68, Mac69**], it is shown that for any integer $g \geq 2$, a Riemann surface of genus g with at least $8(g+1)$ automorphisms exists, and that no larger automorphism groups exist for infinitely many g; specifically, the bound $8(g+1)$ is sharp for all genera of the form $g = 2p+1$ where p is a prime with the properties $p > 863$ and $\gcd((p-1)/2, 42) = 1$. Furthermore, there is always a Riemann surface of genus g with at least $8(g+3)$ automorphisms if g is divisible by 3 or $g \equiv 1 \pmod 4$, and this bound is sharp for all genera of the form $g = p+1$ where p is a prime that is larger than 214 and is congruent to one of 23, 47, and 59 modulo 60.

g	Adm.	Act.		Iso.	Grp.	Orb.	$|G|_{max}$	Max. sign.	q_{max}
2	20	20=	18+ 2	19	21	21	48	$(0;2,3,8)$	48
3	42	41=	34+ 7	37	49	55	168	$(0;2,3,7)$	84
4	61	56=	47+ 9	44	64	73	120	$(0;2,4,5)$	40
5	68	65=	51+ 14	64	93	116	192	$(0;2,3,8)$	48
6	95	75=	63+ 12	59	87	105	150	$(0;2,3,10)$	30
7	114	98=	72+ 26	86	148	208	504	$(0;2,3,7)$	84
8	106	92=	74+ 18	65	108	141	336	$(0;2,3,8)$	48
9	151	135=	102+ 33	154	268	428	320	$(0;2,4,5)$	40
10	182	168=	130+ 38	119	226	335	432	$(0;2,3,8)$	48
11	161	145=	103+ 42	118	232	424	240	$(0;2,4,6)$	24
12	186	167=	128+ 39	98	201	329	120	$(0;2,4,15)$	120/11
13	254	222=	158+ 64	206	453	952	360	$(0;2,3,10)$	30
14	212	183=	136+ 47	99	229	365	1092	$(0;2,3,7)$	84
15	282	254=	178+ 76	176	408	924	504	$(0;2,3,9)$	36
16	326	283=	200+ 83	139	386	789	720	$(0;2,3,8)$	48
17	306	281=	194+ 87	346	733	1834	1344	$(0;2,3,7)$	84
18	300	277=	197+ 80	117	337	742	168	$(0;2,4,21)$	168/17
19	430	398=	272+126	290	791	2119	720	$(0;2,4,5)$	40
20	366	337=	235+102	136	425	936	228	$(0;2,6,6)$	12
21	489	436=	289+147	368	941	3365	480	$(0;2,4,6)$	24
22	510	441=	299+142	187	628	1762	1008	$(0;2,3,8)$	48
23	424	391=	241+150	193	718	2694	192	$(0;2,4,48)$	96/11
24	532	499=	337+162	171	625	1812	216	$(0;2,4,27)$	216/23
25	682	637=	418+219	621	1695	7274	720	$(0;2,3,10)$	30
26	611	542=	354+188	184	715	2058	750	$(0;2,3,10)$	30
27	686	638=	402+236	276	1101	5109	624	$(0;2,3,12)$	24
28	773	731=	477+254	306	1147	4024	1296	$(0;2,3,8)$	48
29	735	689=	423+266	483	1642	9812	672	$(0;2,4,6)$	24
30	786	736=	471+265	187	930	3706	264	$(0;2,4,33)$	264/29
31	1005	921=	567+354	404	1786	10258	720	$(0;2,4,6)$	24
32	851	805=	503+302	189	1048	4404	372	$(0;2,6,6)$	12
33	1007	949=	576+373	1013	2843	18904	768	$(0;3,3,4)$	24
34	1084	1019=	618+401	255	1444	7664	1320	$(0;2,4,5)$	40
35	1059	1013=	596+417	332	1848	13482	544	$(0;2,4,8)$	16
36	1264	1150=	704+446	253	1495	8041	672	$(0;2,3,16)$	96/5
37	1410	1346=	816+530	880	3452	31541	1728	$(0;2,3,8)$	48
38	1215	1140=	672+468	205	1500	8473	444	$(0;2,6,6)$	12
39	1390	1325=	763+562	381	2424	21882	912	$(0;2,3,12)$	24
40	1608	1518=	903+615	341	2192	16148	936	$(0;2,3,12)$	24
41	1580	1520=	875+645	1163	4192	48952	960	$(0;2,4,6)$	24
42	1630	1535=	891+644	244	2000	14259	410	$(0;2,5,10)$	10
43	1894	1805=	1028+777	549	3585	41110	1512	$(0;2,3,9)$	36
44	1733	1670=	954+716	244	2220	17308	516	$(0;2,6,6)$	12
45	2047	1946=	1097+849	788	4193	68873	1320	$(0;2,3,10)$	30
46	2196	2084=	1187+897	436	3211	31616	2160	$(0;2,3,8)$	48
47	2033	1950=	1055+895	401	3638	52414	384	$(0;2,4,96)$	192/23
48	2261	2167=	1221+946	273	2814	28802	408	$(0;2,4,51)$	408/47

TABLE 13. Numbers of Signatures and Characters for Small Genera

CHAPTER 6

Classification for Fixed Group: Real Characters

In this chapter, we start to study the problem of classifying *all* characters of a given group that come from Riemann surfaces. For that, we interpret the properties listed in Section 10 as necessary conditions on a character to come from a Riemann surface.

The easiest case is that of *real* characters that come from Riemann surfaces, because they are in fact rational and their values are determined by fixed point numbers (see Corollary 12.3). Therefore we restrict our attention mainly to real characters. Nonreal irrationalities will be treated in Chapter 7.

Following [**JK81**, p. 11], we call a group G *ambivalent* if every element in G is G-conjugate to its inverse. A group is ambivalent if and only if all of its characters are real, so in this chapter, we are particularly interested in ambivalent groups.

21. Example: $G = D_{2p}$

The aim of this section is to prove

THEOREM 21.1. *Let $G = D_{2p}$ be the dihedral group of order $2p$, where p is an odd prime. A character χ of G comes from a Riemann surface if and only if χ is a rational character of degree at least 2 and with the property that $\chi(\sigma) \leq 1$ for all $\sigma \in G^{\times}$.*

A presentation for G is $\langle a, b \mid a^p, b^2, (ab)^2 \rangle$, the group has $2 + (p-1)/2$ conjugacy classes, namely $\{1\}$, $\{a^i, a^{-i}\}$ for $1 \leq i \leq (p-1)/2$, and $\{a^i b \mid 0 \leq i \leq p-1\}$. We see that G is ambivalent, and each character that comes from a Riemann surface is rational. So we need to consider only the rational irreducible characters of G as constituents; they are shown in Table 14.

Let $\chi = a_1 1_G + a_2 \chi_2 + a_3 \chi_3$, with integers a_1, a_2, a_3. Then the conditions that $\chi(1) \geq 2$ and $\chi(\sigma) \leq 1$ for all $\sigma \in G^{\times}$ are equivalent to the system of

	1	a	b
1_G	1	1	1
χ_2	1	1	-1
χ_3	$p-1$	-1	0

TABLE 14. Rational Irreducible Characters of D_{2p}

inequalities

$$a_1, \quad a_2, \quad a_3 \ \geq \ 0,$$
$$a_1 + a_2 - a_3 \ \leq \ 1,$$
$$a_1 - a_2 \ \leq \ 1, \quad \text{and}$$
$$a_1 + a_2 + (p-1)a_3 \ \geq \ 2 \ .$$

By Corollary 12.3, this condition is clearly necessary for χ to come from a Riemann surface. So we have to show that each such choice of coefficients belongs to a character that comes from a Riemann surface, and we do this by constructing a surface kernel epimorphism for each such character.

For example, consider the surface kernel epimorphism $\Phi_0 \colon \Gamma(0; 2, 2, p, p) \to G$, defined by

$$c_1 \mapsto b, \quad c_2 \mapsto b, \quad c_3 \mapsto a, \quad c_4 \mapsto a^{-1}.$$

Let $X = \mathcal{U}/\ker(\Phi_0)$ be the associated Riemann surface. Then

$$g(X) = 1 - 2p + (2p/2) \cdot (1 - 1/2 + 1 - 1/2 + 1 - 1/p + 1 - 1/p) = p - 1.$$

Since $\mathrm{Tr}(\Phi_0)$ is faithful and has degree $p-1$, we must have $\mathrm{Tr}(\Phi_0) = \chi_3$. But let us compute the character values from Φ_0. As we have seen in Example 10.6, $|\mathrm{Fix}_X(a)| = 4$ and $|\mathrm{Fix}_X(b)| = 2$. Using Corollary 12.3, we get

$$
\begin{array}{lclcl}
\mathrm{Tr}(\Phi_0)(1) & = & g(X) & = & p-1, \\
\mathrm{Tr}(\Phi_0)(a) & = & 1 - \tfrac{1}{2} \cdot |\mathrm{Fix}_X(a)| & = & -1, \\
\mathrm{Tr}(\Phi_0)(b) & = & 1 - \tfrac{1}{2} \cdot |\mathrm{Fix}_X(b)| & = & 0,
\end{array}
$$

so indeed $\mathrm{Tr}(\Phi_0) = \chi_3$.

Once we know a surface kernel homomorphism $\Phi \colon \Gamma \to G$, we can easily construct others, as we have seen in Example 12.6. There we increased the orbit genus of Γ by 1, and obtained the character $\mathrm{Tr}(\Phi) + \rho_G$ for the new epimorphism. Let us call the transformation $\psi \mapsto \psi + \rho_G$ of characters T_0.

Now we generalize this idea. Specifically, we can replace $\Gamma(g_0; m_1, m_2, \ldots, m_r)$ by a group with $r + 1$ instead of r periods, and replace an elliptic generator of order p with image a^i by two generators of order p with images a^{-i} and a^{2i}, respectively. We call a transformation of this kind T_3. By Lemma 10.4, it increases the number of fixed points of a by 2, leaves the number of fixed points of b unchanged, and increases the orbit genus of the kernel by $(2p/2) \cdot (1 - 1/p) = p - 1$. The character of the new epimorphism is thus $T_3(\psi) = \psi + \chi_3$.

Another possibility is to replace one image a^{2i} by the images b and $a^{-i}ba^i$, using that $ab = ba^{-1}$. This transformation, called T_2, decreases the number of fixed points of a by 2, increases the number of fixed points of b by 2, changes the orbit genus of the kernel from g to $g - (2p/2) \cdot (1 - 1/p) + (2p/2) \cdot (1 - 1/2 + 1 - 1/2) = g + 1$. For the character, we get $T_2(\psi) = \psi + \chi_2$.

By means of the inequalities above, we can represent the G-characters that come from Riemann surfaces (which we will call the *interesting* characters below) as points with integral coordinates in a 3-dimensional cone. The layers for $a_1 \in \{0, 1, 2\}$ are shown in Figure 1. The circled points correspond to the characters $\mathrm{Tr}(\Phi_0)$, $T_0(\mathrm{Tr}(\Phi_0))$, and $T_0^2(\mathrm{Tr}(\Phi_0))$, and the line $a_1 + a_2 + (p - 1)a_3 = 2$ is drawn for $p = 3$ in the figure; note that the lines for all odd primes p exclude the same points with integral coordinates.

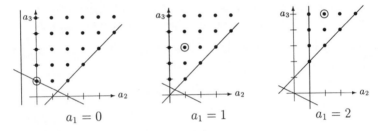

$a_1 = 0$ $a_1 = 1$ $a_1 = 2$

FIGURE 1. Realizable Characters for D_{2p}, with $a_1 \leq 2$.

Each application of T_3 leads from a point to its upper neighbor, T_2 leads to the neighbor on the right, and T_0 to the next layer.

Figure 1 shows that T_2 cannot be applied arbitrarily often. This is clear because T_2 needs an elliptic generator with image a^i. We can apply T_2 at most twice to $\mathrm{Tr}(\Phi_0)$, at most three times to $T_3(\mathrm{Tr}(\Phi_0))$, and so on. In other words, $T_2^{\alpha_2} T_3^{\alpha_3}(\mathrm{Tr}(\Phi_0)) = \alpha_2\chi_2 + (1 + \alpha_3)\chi_3$ is admissible if $\alpha_2 \leq 2 + \alpha_3$. This is sufficient to establish all interesting characters with $a_1 = 0$.

For the layers with $a_1 > 0$, it turns out that $T_0(\mathrm{Tr}(\Phi_0))$ together with its images under T_2 and T_3 does not cover all characters we need. With the same argument as for $a_1 = 0$, it suffices to construct surface kernel epimorphisms for the characters $1_G + \chi_3$ and $2 \cdot 1_G + \chi_2 + 2 \cdot \chi_3 = 1_G + \rho_G$ to establish all interesting characters with $a_1 = 1$ and $a_1 = 2$, respectively. The latter character comes from a Riemann surface by Example 12.5. If the former is afforded by a surface kernel epimorphism $\Phi \colon \Gamma \to G$ then let $X = \mathcal{U}/\ker(\Phi)$. From $|\mathrm{Fix}_X(a)| = 2(1 - \chi(a)) = 2$ and $|\mathrm{Fix}_X(b)| = 2(1 - \chi(b)) = 0$, we see that the signature of Γ is $(1; p)$. We define Φ by mapping the two hyperbolic generators to a and b, respectively, and mapping the unique elliptic generator to a^2; then Φ is a homomorphism because $a^2[a, b]$ is the identity in G.

Finally, if $a_1 > 2$, all interesting characters are described by the two inequalities $a_1 + a_2 - a_3 \leq 1$ and $a_1 - a_2 \leq 1$, and for each such character χ, $\chi - \rho_G$ also satisfies the system of inequalities. By induction on a_1, we may assume that $\chi - \rho_G$ comes from a Riemann surface, so $\chi = T_0(\chi - \rho_G)$ comes from a Riemann surface, too.

REMARK 21.2. By the fact that χ_3 is the unique faithful rational irreducible character of D_{2p}, it must be a constituent of *every* character of D_{2p} that comes from a Riemann surface. Thus $p - 1$ is the minimal genus larger than or equal to 2 for a Riemann surface that has D_{2p} as group of automorphisms. But note that each dihedral group is an automorphism group of a Riemann surface of genus 0, as it is an epimorphic image of $\Gamma(0; 2, 2, p)$, see Section 3.4, so the strong symmetric genus (see Section 16) of each dihedral group is 0.

22. Necessary Conditions on $\chi + \bar{\chi}$

The properties of characters coming from Riemann surfaces that we used in the preceding section are derived from properties of certain sets of fixed points, see Section 10. Now we treat them more systematically.

22.1. The Conditions (E) and (RH).

DEFINITION 22.1 (see for example [**Kur84, KK91**]). Let G be a finite group, and χ a class function of G.

We say that χ satisfies the condition (E) if and only if $\chi(1) \geq 2$, and $\chi(h) + \overline{\chi(h)}$ is an integer for all $h \in G$.

Suppose that χ satisfies the condition (E). For each $H = \langle h \rangle \in CY(G)$, the values

$$r_\chi(h) = r_\chi(H) = 2 - (\chi(h) + \overline{\chi(h)}),$$

$$r_\chi^*(h) = r_\chi^*(H) = r_\chi(H) - \sum_{K \in CY(G, H)} r_\chi^*(K), \quad \text{and}$$

$$l_\chi(h) = l_\chi(H) = r_\chi^*(H)/[N_G(H){:}H]$$

are well-defined. We say that χ is an (RH)-class function (or that χ satisfies the condition (RH)) if and only if χ satisfies the condition (E) and $l_\chi(H)$ is a nonnegative integer for all $H \in CY(G)$.

The maps r_χ, r_χ^*, and l_χ are class functions on $G^\times$, they are invariants of $\chi + \overline{\chi}$ rather than of χ.

A character χ of G that comes from a Riemann surface X satisfies the condition (E) by Corollary 12.3. From this, we get $r_\chi(H) = |\text{Fix}_X(H)|$, so Lemma 10.2 yields $r_\chi^*(H) = |\text{Fix}_X^G(H)|$. If π denotes the natural projection $X \to X/G$ then Lemma 10.3 yields $l_\chi(H) = |\pi(\text{Fix}_X^G(H))|$, the number of fibres with point stabilizer in G conjugate to H, hence χ satisfies the condition (RH).

REMARK 22.2. We could have replaced condition (E) by the stronger condition that $r_\chi(h)$ is a *nonnegative* integer. But the resulting inequalities would have been consequences of the inequalities that we get by condition (RH). Note that if $r_\chi^*(H) \geq 0$ for all $H \in CY(G)$ holds then $r_\chi(H) = r_\chi^*(H) + \sum_{K \in CY(G,H)} r_\chi^*(K)$ is also nonnegative.

EXAMPLE 22.3. We reformulate the result of Theorem 21.1. Let p be an odd prime, and χ a character of D_{2p} with $\chi(1) \geq 2$. Then $l_\chi(a) = r_\chi(a)/2 = 1 - \chi(a)$ and $l_\chi(b) = r_\chi(b) = 2 - 2\chi(b)$ are nonnegative integers if and only if $\chi(a)$ and $\chi(b)$ are integers ≤ 1. So χ comes from a Riemann surface if and only if it satisfies the conditions (E) and (RH). $\diamond$

22.2. Permutation Characters. Now let us look at the condition (RH) from a different viewpoint. Suppose that the character χ of the group G comes from the Riemann surface X, and define

$$X_0 = \biguplus_{H \in CY(G)} \text{Fix}_X^G(H),$$

the points in X with nontrivial stabilizer in G. Then

$$|X_0| = \sum_{H \in CY(G)/\sim_G} [G{:}N_G(H)] \cdot |\text{Fix}_X^G(H)| = \sum_{H \in CY(G)/\sim_G} l_\chi(H) \cdot \frac{|G|}{|H|},$$

and G acts on X_0 as a permutation group. If we set $r_\chi(1) = |X_0|$ then r_χ is the permutation character of this action (see Section 6). It can be decomposed into a sum of transitive permutation characters where the multiplicity of 1_H^G equals the number of fibres of the covering $X \to X/G$ with point stabilizer in G conjugate to H, see Lemma 10.3. In other words,

$$r_\chi = \sum_{H \in CY(G)/\sim_G} l_\chi(H) \cdot 1_H^G.$$

In particular, the scalar product $[r_\chi, 1_G]$ equals $\sum_{H \in CY(G)/\sim_G} l_\chi(H)$, the number of fibres in X_0.

The following lemma generalizes this relation. Note that we do *not* assume that χ comes from a Riemann surface, we do not even assume that χ is a character or that it satisfies the condition (RH).

LEMMA 22.4. *Let χ be a class function of G that satisfies the condition* (E), *and set*

$$r_\chi(1) = \sum_{H \in CY(G)/\sim_G} l_\chi(H) \cdot \frac{|G|}{|H|}.$$

Then

$$r_\chi = \sum_{H \in CY(G)/\sim_G} l_\chi(H) \cdot 1_H^G.$$

Proof. For the identity element, the equality is obvious by the definition of $r_\chi(1)$, so let $h \in G^\times$. Consider the identity

$$\biguplus_{\substack{\sigma \in G \\ H \in CY(G) \\ h^\sigma \in H}} \{(\sigma, H)\} \;=\; \{(\sigma, H) \mid H \in CY(G), \sigma \in G, h^\sigma \in H\}$$

$$=\; \biguplus_{H \in CY(G)} \;\biguplus_{\substack{\sigma \in G \\ h^\sigma \in H}} \{(\sigma, H)\}.$$

Taking cardinalities on both sides and mapping each set $\{(\sigma, H)\}$ to $r_\chi^*(H)$ yields

$$\sum_{\sigma \in G} \sum_{\substack{H \in CY(G) \\ h^\sigma \in H}} r_\chi^*(H) = \sum_{H \in CY(G)} \sum_{\substack{\sigma \in G \\ h^\sigma \in H}} r_\chi^*(H).$$

By definition, the inner sum on the left hand side is equal to $r_\chi(h^\sigma) = r_\chi(h)$, independently of σ. The number of summands in the inner sum on the right hand side is equal to $|\{\sigma \in G \mid h^\sigma \in H\}| = |h^G \cap H| \cdot |C_G(h)|$, so we have

$$|G| \cdot r_\chi(h) = \sum_{H \in CY(G)} |h^G \cap H| \cdot |C_G(h)| \cdot r_\chi^*(H),$$

which yields

$$
\begin{aligned}
r_\chi(h) &= \sum_{H \in CY(G)} \frac{|h^G \cap H|}{|h^G|} \cdot r_\chi^*(H) = \sum_{H \in CY(G)} \frac{|H|}{|G|} \cdot 1_H^G(h) \cdot r_\chi^*(H) \\
&= \sum_{H \in CY(G)/\sim_G} \frac{|G|}{|N_G(H)|} \cdot \frac{|H|}{|G|} \cdot 1_H^G(h) \cdot r_\chi^*(H) \\
&= \sum_{H \in CY(G)/\sim_G} l_\chi(H) \cdot 1_H^G(h),
\end{aligned}
$$

as desired. $\square$

Since the class functions $\chi + \overline{\chi}$ and $2 \cdot 1_G - r_\chi$ are equal on $G^\times$, the difference is a multiple of the regular character ρ_G. Counting the multiplicities of the trivial character in both expressions, we get

$$\chi + \overline{\chi} = 2 \cdot 1_G - r_\chi + (2 \cdot [\chi, 1_G] - 2 + [r_\chi, 1_G]) \cdot \rho_G.$$

An alternative formulation is obtained using the fact that $\rho_G = (\rho_H)^G$ for each subgroup of G.

COROLLARY 22.5. Let χ be a G-character that satisfies the condition (E), and $g_0 = [\chi, 1_G]$. Then

$$\chi + \overline{\chi} = 2 \cdot 1_G + 2(g_0 - 1) \cdot \rho_G + \sum_{H \in CY(G)/\sim_G} l_\chi(H) \cdot (\rho_H - 1_H)^G.$$

Evaluated at the identity element, this reads as follows (see [**Kur84**, Prop. 1]).

COROLLARY 22.6. Under the conditions of Corollary 22.5, we have

$$\chi(1) - 1 = |G| (g_0 - 1) + \frac{|G|}{2} \cdot \sum_{H \in CY(G)/\sim_G} l_\chi(H) \cdot (1 - 1/|H|).$$

If χ is a character that comes from the Riemann surface X then this equation is just the known Riemann–Hurwitz Formula of Lemma 3.13. The fact that we can prove it also if χ does not come from a Riemann surface means that the Riemann–Hurwitz Formula does not impose more than the condition (RH), which justifies the name for this condition. Moreover, also the representability of r_χ as a sum of transitive permutation characters as in Corollary 22.5 is not a stronger condition than (RH). In fact these two conditions are equivalent, because $\mathrm{Rat}(G)$ is linearly independent over $\mathbb{Q}$ and spans the same $\mathbb{Z}$-lattice as

$$\{1_H^G \mid H \in CY(G)/\sim_G\} \cup \{\rho_G\},$$

by Artin's Induction Theorem [**Isa76**, Theorem 5.21]. Hence

$$\{(\rho_H - 1_H)^G \mid H \in CY(G)/\sim_G\} \cup \{\rho_G\}$$

is also linearly independent.

REMARK 22.7. An (RH)-class function of G is not necessarily a character. For example, take $G = D_{2p}$ as in Section 21, and consider

$$\chi = -\chi_2 + 2\chi_3 = 1_G - \rho_G + 2 \cdot (\rho_G - 1_{\langle a \rangle}^G).$$

The corresponding point in Figure 1 has negative a_2 coordinate but is not excluded by the condition (RH).

EXAMPLE 22.8. We cannot expect that r_χ is twice a character (or, equivalently, that all $l_\chi(H)$ are even), even if χ is a rational (RH)-character. As an example, consider the regular character $\chi = \rho_G$ of the Klein four group $G = \langle a_1, a_2 \mid a_1^2, a_2^2, (a_1 a_2)^2 \rangle$. In Section 25, we will show that χ comes from a Riemann surface.

We have $l_\chi(a) = r_\chi^*(a)/2 = r_\chi(a)/2 = 1 - \chi(a) = 1$ for all $a \in G^\times$, so $r_\chi = 1_{\langle a_1 \rangle}^G + 1_{\langle a_2 \rangle}^G + 1_{\langle a_1 a_2 \rangle}^G = \rho_G + 2 \cdot 1_G$, and hence $2 \cdot \chi + r_\chi = 2 \cdot 1_G + 3 \cdot \rho_G$. $\diamond$

EXAMPLE 22.9 (cf. [**Kur84**, Proposition 4]). We show that the restriction of an (RH)-character χ of G to a subgroup $U \leq G$ is an (RH)-character of U.

Let χ be a G-character that satisfies the condition (E), and

$$\chi + \overline{\chi} - 2 \cdot 1_G = 2(g_0 - 1) \cdot \rho_G + \sum_{H \in CY(G)/\sim_G} c_H \cdot \left(\rho_G - 1_H^G \right).$$

Then $g_0 = [\chi, 1_G]$, and $c_H = l_\chi(H)$ for all $H \in CY(G)/\sim_G$. For the restriction χ_U of χ to U, we get

$$\chi_U + \overline{\chi}_U - 2 \cdot 1_U$$
$$= 2(g_0 - 1)[G{:}U] \cdot \rho_U + \sum_{H \in CY(G)/\sim_G} c_H \cdot \left([G{:}U] \cdot \rho_U - (1_H^G)_U \right).$$

By Mackey's Theorem (see page 24), the characters of the form $(1_H^G)_U$ can be decomposed into a sum of at most $[G{:}U]$ permutation characters 1_K^U, with $K \in CY(U)$. So we may reorder the summation according to conjugacy classes of cyclic subgroups of U to get

$$\chi_U + \overline{\chi}_U - 2 \cdot 1_U = 2(g_0' - 1) \cdot \rho_U + \sum_{K \in CY(U)/\sim_U} c_K' \cdot \left(\rho_U - 1_K^U \right),$$

where the coefficients c_K' are sums of multiples of the c_H, and $2(g_0' - 1)$ is a sum of $2(g_0 - 1)[G{:}U]$ and multiples of the c_H. Note that the c_K' are independent of the choices of representatives in $CY(G)$ and $CY(U)$.

Now the claim follows, because the c_K' are nonnegative integers if the c_H are, and $g_0' = [\chi_U, 1_U]$ is also a nonnegative integer. $\diamond$

22.3. Compatibility of Characters and Homomorphisms. Suppose that the character χ satisfies the condition (RH). Following Lemma 10.4, we can try to interpret the value $l_\chi(H)$ as the number of those elliptic generators of a group Γ as in Theorem 3.2 that are mapped to generators of G-conjugates of H under an appropriate homomorphism $\Phi \colon \Gamma \to G$.

If χ comes from a Riemann surface X then certainly such a homomorphism exists, namely a surface kernel epimorphism $\Phi \colon \Gamma \to G$ with kernel K such that $\mathcal{U}/K \cong X$. Using the terminology introduced in Definition 13.1, we have $\chi = \mathrm{Tr}(\Phi)$ then, and $\Gamma = \Gamma(\chi)$.

Otherwise it is not clear that such a homomorphism exists at all, and if it does whether a surjective one exists. So we need some notion of the relation between characters and homomorphisms.

DEFINITION 22.10. For an (RH)-character χ of G, let $r = \sum_{H \in CY(G)/\sim_G} l_\chi(H)$. Define $L(\chi) = (L_1, L_2, \ldots, L_r)$, where each L_i is the rational class of an element in $G^\times$, and where the rational class of σ occurs exactly $l_\chi(\langle \sigma \rangle)$ times.

Obviously, a G-character χ comes from a Riemann surface if and only if there is a $\Phi \in \text{Epi}_{L(\chi)}(\Gamma(\chi), G)$ such that $\text{Tr}(\Phi) = \chi$, and clearly every surface kernel epimorphism Φ with the property $\text{Tr}(\Phi) = \chi$ satisfies $\Phi \in \text{Epi}_{L(\chi)}(\Gamma(\chi), G)$. For the converse, the following holds. Note that we get a relation only for $\chi + \overline{\chi}$, not for χ itself.

LEMMA 22.11. *Let χ be an (RH)-character of G, and $\Phi \in \text{Hom}_{L(\chi)}(\Gamma(\chi), G)$ with $\Phi(\Gamma(\chi)) = U \leq G$. Then*

$$\left(\text{Tr}(\Phi) + \overline{\text{Tr}(\Phi)} - 2 \cdot 1_U \right)^G = \chi + \overline{\chi} - 2 \cdot 1_G.$$

Proof. The condition $\Phi \in \text{Hom}_{L(\chi)}(\Gamma(\chi), G)$ means that the orbit genus $g_0 = [\chi, 1_G]$ of $\Gamma(\chi)$ is equal to $[\text{Tr}(\Phi), 1_U]$, and

$$l_\chi(H) = \sum_{\substack{K \in CY(U)/\sim_U \\ K \sim_G H}} l_{\text{Tr}(\Phi)}(K)$$

for all $H \in CY(G)$. We may choose each representative $H \in CY(G)/\sim_G$ inside U if H is G-conjugate to a subgroup of U, and the character $(1_H)^U$ is well-defined. Since $l_\chi(H) = 0$ if H is not G-conjugate to a subgroup of U, we get

$$\sum_{\substack{H \in CY(G)/\sim_G \\ l_\chi(H) \neq 0}} l_\chi(H) \cdot (\rho_U - 1_H^U) = \sum_{K \in CY(U)/\sim_U} l_{\text{Tr}(\Phi)}(K) \cdot (\rho_U - 1_K^U),$$

and the statement follows from Corollary 22.5, which yields

$$\chi + \overline{\chi} - 2 \cdot 1_G = 2(g_0 - 1) \cdot \rho_G + \sum_{H \in CY(G)/\sim_G} l_\chi(H) \cdot (\rho_G - 1_H^G)$$

and

$$\text{Tr}(\Phi) + \overline{\text{Tr}(\Phi)} - 2 \cdot 1_U = 2(g_0 - 1) \cdot \rho_U + \sum_{K \in CY(U)/\sim_U} l_{\text{Tr}(\Phi)}(K) \cdot (\rho_U - 1_K^U),$$

and from the transitivity of induction.   □

EXAMPLE 22.12. Let G be the alternating group A_5, with character table as in [CCN+85, p. 2], and consider the character $\chi = \chi_1 + \chi_2 + \chi_3 + \chi_4 + \chi_5$ of G. We have $\chi(1A) = 16$, $\chi(2A) = 0$, and $\chi(3A) = \chi(5A) = \chi(5B) = 1$, thus $l_\chi(2A) = 1$ and $l_\chi(3A) = l_\chi(5A) = l_\chi(5B) = 0$. So χ satisfies the condition (RH), and $\Gamma(\chi) = \Gamma(1; 2)$. But χ cannot come from a Riemann surface, because each pair of elements in A_5 whose commutator is an involution turns out to generate a proper subgroup $U \cong A_4$. Following Lemma 22.11, we have

$\chi - 1_G = (\psi - 1_U)^G$ for a character ψ of degree 4 of U that comes from a Riemann surface; moreover $\psi - 1_U$ is the unique faithful irreducible character of U.

A generalization of this example can be found in Section 27.2.   ◇

EXAMPLE 22.13. As above, take $G = A_5$, and consider surface kernel homomorphisms from $\Gamma = \Gamma(0; 5, 5, 5)$ to G. Since G is generated by the permutations $\sigma = (1, 2, 3, 4, 5)$, $(1, 4, 5, 2, 3)$, and $(1, 2, 4, 5, 3)$, whose product is the identity, there is a surface kernel epimorphism $\Phi_1 \colon \Gamma \to G$. We set $\chi = \mathrm{Tr}(\Phi_1)$, and get $\chi(1A) = 13$, $\chi(2A) = \chi(3A) = 1$, and $\chi(5A) = \chi(5B) = -2$, and thus $\chi = 2 \cdot \chi_4 + \chi_5$.

As a second homomorphism, Φ_2 is defined to map the three generators of Γ to σ, σ, σ^3. The image of Φ_2 is the cyclic group $U = \langle \sigma \rangle$ of order 5. Let ε denote the unique irreducible character of U with $\varepsilon(\sigma) = \zeta_5$. Then $\mathrm{Tr}(\Phi_2) = \varepsilon + \varepsilon^2$. We have $(\mathrm{Tr}(\Phi_2) - 1_U)^G = \chi - 1_G$, a relation that is even stronger than the one guaranteed by Lemma 22.11. (Note that $\mathrm{Tr}(\Phi_2)$ is not real.)   ◇

The notion $\chi \in \mathrm{Hom}_{L(\chi)}(\Gamma(\chi), G)$ implies that

$$r_\chi^*(H) = [N_G(H){:}H] \cdot |\{i \mid 1 \le i \le r, H \sim_G \langle \Phi(c_i) \rangle\}|,$$

which is nothing but the second statement of Lemma 10.4, in a formulation that does not involve Riemann surfaces. The first statement of this lemma is a consequence of the second, so we get the following generalization.

LEMMA 22.14. *Let χ be a G-character, and $\Phi \in \mathrm{Hom}_{L(\chi)}(\Gamma(\chi), G)$. Then*

$$r_\chi(H) = |N_G(H)| \cdot \sum_{\substack{1 \le i \le r \\ m | m_i \\ H \sim_G \langle \Phi(c_i)^{m_i/m} \rangle}} \frac{1}{m_i}$$

for all $H \in CY(G)$.

Proof. Let $\widehat{CY}(G, H) = CY(G, H) \cup \{H\}$. Then

$$r_\chi(H) = \sum_{K \in \widehat{CY}(G,H)} r_\chi^*(K) = \sum_{K \in \widehat{CY}(G,H)/\sim_G} [N_G(H){:}N_G(K)] \cdot r_\chi^*(K),$$

because H is contained in exactly $[G{:}N_G(K)]/[G{:}N_G(H)]$ conjugates of $K \in \widehat{CY}(G, H)$. Inserting the above equality for r_χ^*, we get

$$r_\chi(H) = |N_G(H)| \cdot \sum_{K \in \widehat{CY}(G,H)/\sim_G} \frac{1}{|K|} \cdot |\{i \mid 1 \le i \le r, K \sim_G \langle \Phi(c_i) \rangle\}|,$$

from which the formula in the lemma follows.   □

REMARK 22.15. Alternatively, we could prove Lemma 22.14 using the equality

$$\left(r_{\mathrm{Tr}(\Phi)}\right)^G = r_\chi$$

of class functions on $G^\times$, which is just a reformulation of Lemma 22.11. (If we define $r_\chi(1)$ and $r_{\mathrm{Tr}(\Phi)}(1)$ as in Lemma 22.4 then this equality holds also for the degrees.)

22.4. Linearity of the Condition (RH). Let χ be a class function of G that satisfies the condition (E). The statement of Corollary 22.5 can be expressed by the following linear relation between the values $l_\chi(H)$ and the coefficients of the decomposition of $\chi + \overline{\chi}$ into rational irreducible characters.

LEMMA 22.16. *Let H_0 denote the trivial subgroup of G, $\widehat{CY}(G) = CY(G) \cup \{H_0\}$, and $l_\chi(H_0) = [\chi, 1_G]$ for each class function χ of G. Furthermore, set $\psi_{G,H_0} = 2 \cdot \rho_G$ and $\psi_{G,H} = \rho_G - 1_H^G$ for $H \in CY(G)$.*

Define the vector v and the matrix T by

$$v = (v_\varphi)_{\varphi \in \mathrm{Rat}(G)}, \quad \text{with} \quad v_{1_G} = 0 \quad \text{and} \quad v_\varphi = \frac{-2 \cdot \varphi(1)}{[\varphi, \varphi]} \quad \text{otherwise},$$

and

$$T = (T_{H,\varphi})_{H \in \widehat{CY}(G)/\sim_G, \varphi \in \mathrm{Rat}(G)}, \quad \text{with} \quad T_{H,\varphi} = \frac{[\psi_{G,H}, \varphi]}{[\varphi, \varphi]}.$$

If χ is a class function of G with $\chi + \overline{\chi} = \sum_{\varphi \in \mathrm{Rat}(G)} a_\varphi \cdot \varphi$ then

$$(a_\varphi)_{\varphi \in \mathrm{Rat}(G)} = v + (l_\chi(H))_{H \in \widehat{CY}(G)/\sim_G} \cdot T.$$

Proof. With $g_0 = [\chi, 1_G]$, we have

$$
\begin{aligned}
a_\varphi \cdot [\varphi, \varphi] &= [\chi + \overline{\chi}, \varphi] \\
&= [2 \cdot 1_G + 2(g_0 - 1) \cdot \rho_G, \varphi] + \sum_{H \in CY(G)/\sim_G} l_\chi(H) \cdot [(\rho_H - 1_H)^G, \varphi] \\
&= [2 \cdot 1_G - 2 \cdot \rho_G, \varphi] + \sum_{H \in \widehat{CY}(G)/\sim_G} l_\chi(H) \cdot [\psi_{G,H}, \varphi] \\
&= v_\varphi \cdot [\varphi, \varphi] + \sum_{H \in \widehat{CY}(G)/\sim_G} l_\chi(H) \cdot [\psi_{G,H}, \varphi].
\end{aligned}
$$

$\square$

REMARK 22.17. Note that v and T are independent of χ, that v consists of nonpositive even integers, and the entries of T are nonnegative integers because $\psi_{G,H}$ contains each irreducible constituent of φ with the same multiplicity. Moreover, we have $v_\varphi = -T_{H_0,\varphi}$ for $\varphi \neq 1_G$; that is, except v_{1_G}, v is the negative of the H_0-row of T.

From the statements after Corollary 22.6, it is clear that T is invertible.

EXAMPLE 22.18. Let $\varphi \in \operatorname{Rat}(G)$ and $H \in \widehat{CY}(G)$. Clearly $T_{H,\varphi} \neq 0$ if $|H| = 1$, and for $H \in CY(G)$ we get

$$T_{H,\varphi} \cdot [\varphi, \varphi] = [\psi_{G,H}, \varphi] = [\rho_G, \varphi] - [1_H^G, \varphi] = \varphi(1) - [1_H, \varphi_H]$$

by Frobenius reciprocity. Thus $T_{H,\varphi} = 0$ if and only if $|H| > 1$ and $H \subseteq \ker(\varphi)$.   ◇

An immediate consequence is

LEMMA 22.19. Let χ be an (RH)-character of G for which $[\chi, 1_G] = 0$ holds. Then the image of each $\Phi \in \operatorname{Hom}_{L(\chi)}(\Gamma(\chi), G)$ is not contained in a proper normal subgroup of G.

Proof. Suppose that N is a proper normal subgroup of G such that the image of Φ is contained in N. Then $l_\chi(H) = 0$ for all $H \in CY(G)$ with $H \not\subseteq N$. Take a nontrivial character $\varphi \in \operatorname{Rat}(G)$ such that $N \in \ker(\varphi)$; the multiplicity a_φ of φ as a constituent of χ is equal to the negative integer v_φ by Lemma 22.16 and Example 22.18, so χ is not a character.

Alternatively, we can prove this fact (in an even more elementary way) as follows. If $l_\chi(H) = 0$ for all $H \in CY(G)$ with $H \not\subseteq N$ then $r_\chi(H) = 0$ for such H, and hence $\chi(\sigma) = 1$ for each $\sigma \in G \setminus N$. But this means that $|N| \cdot ([\chi_N, 1_N] - 1) = |G| \cdot ([\chi, 1_G] - 1)$, which forces the scalar product $[\chi_N, 1_N]$ to be negative.   □

EXAMPLE 22.20. As another application of Lemma 22.16, we consider the columns of T that belong to *linear* characters of G. More generally, let $\varphi \in \operatorname{Rat}(G)$ be the Galois orbit sum of the linear character $\vartheta \in \operatorname{Irr}(G)$. Then $[\psi_{G,H}, \varphi] = 2 \cdot [\rho_G, \varphi] = 2 \cdot \varphi(1)$ if $|H| = 1$ and $[\psi_{G,H}, \varphi] = 0$ if $H \subseteq \ker(\varphi)$ by Example 22.18. For $H \not\subseteq \ker(\varphi)$, we get analogously

$$[\psi_{G,H}, \varphi] = \varphi(1) - [1_H^G, \varphi] = \varphi(1) - [1_H, \varphi_H] = \varphi(1).$$

Because $[\varphi, \varphi] = \varphi(1)$, this means that for our choice of φ,

$$T_{H,\varphi} = \begin{cases} 2, & |H| = 1, \\ 0, & |H| \neq 1, H \subseteq \ker(\varphi), \\ 1, & H \not\subseteq \ker(\varphi). \end{cases}$$

In particular $T_{H,\varphi}$ is odd if and only if $H \not\subseteq \ker(\varphi)$.   ◇

EXAMPLE 22.21. Finally, let us look at the example of Section 21 again. Let $G = D_{2p}$. The vector v and the matrix T of Lemma 22.16 are

$$v = (0, -2, -4) \quad \text{and} \quad T = \begin{pmatrix} 2 & 2 & 4 \\ 0 & 0 & 2 \\ 0 & 1 & 1 \end{pmatrix},$$

where the columns are indexed by the characters 1_G, χ_2, and χ_3 as given in Table 14 on page 93, and the rows of T are indexed by the subgroups spanned by 1, a, and b. The values of v and the entries in the first two columns of T are clear from the definition or from Example 22.20. For the computation of the values in the last column of T, we compute $1^G_{\langle a \rangle} = 1_G + \chi_2$ and $1^G_{\langle b \rangle} = 1_G + \chi_3$, from which we get $[\rho_G - 1^G_{\langle a \rangle}, \chi_3] = 2 \cdot [\chi_3, \chi_3]$ and $[\rho_G - 1^G_{\langle b \rangle}, \chi_3] = [\chi_3, \chi_3]$.

Let $\chi = a_1 1_G + a_2 \chi_2 + a_3 \chi_3$, then the relation of Lemma 22.16 reads as

$$(2a_1, 2a_2, 2a_3) = v + (g_0, l_\chi(a), l_\chi(b)) \cdot T,$$

and the six inequalities on page 93 are expressed in terms of the $l_\chi(H)$ as follows.

$$
\begin{aligned}
g_0 &\in \mathbb{Z}^{\geq 0}, \\
-2 + 2g_0 + l_\chi(b) &\in (2\mathbb{Z})^{\geq 0}, \\
-4 + 4g_0 + 2l_\chi(a) + l_\chi(b) &\in (2\mathbb{Z})^{\geq 0}, \\
l_\chi(b) &\in (2\mathbb{Z})^{\geq 0}, \\
l_\chi(a) &\in \mathbb{Z}^{\geq 0}, \quad \text{and} \\
2p \cdot g_0 + (p-1)l_\chi(a) + p \cdot l_\chi(b)/2 &\geq 2p + 1.
\end{aligned}
$$

Figure 2 shows the solutions for $0 \leq g_0 \leq 2$. The circled points correspond to the characters $\mathrm{Tr}(\Phi_0)$, $T_0(\mathrm{Tr}(\Phi_0))$, and $T_0^2(\mathrm{Tr}(\Phi_0))$ introduced in Section 21.

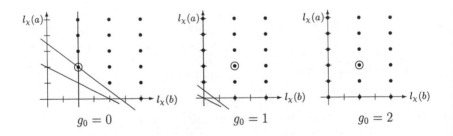

$$g_0 = 0 \qquad\qquad g_0 = 1 \qquad\qquad g_0 = 2$$

FIGURE 2. Realizable Characters for D_{2p}, with $g_0 \leq 2$.

The transformation T_3 of Section 21 increases $l_\chi(a)$ by 1 and leaves $l_\chi(b)$ unchanged, the transformation T_2 decreases $l_\chi(a)$ by 1 and increases $l_\chi(b)$ by 2. So it appears to be more suitable from the viewpoint of Figure 2 to replace T_2 by a transformation that introduces two new generators of order 2 and thus leaves $l_\chi(a)$ unchanged and increases $l_\chi(b)$ by 2. Like T_3, this transformation can be applied arbitrarily often. Its effect on the character is to add $\chi_2 + \chi_3$.

◇

23. Finiteness Results

In this section, we discuss how far the condition (RH) controls the property of characters of coming from Riemann surfaces, that is, how many characters satisfy the condition (RH) but do not come from a Riemann surface.

Given an (RH)-character χ of G, we can split the question whether χ comes from a Riemann surface into three parts, namely whether $\mathrm{Hom}_{L(\chi)}(\Gamma(\chi), G)$ is nonempty, if yes whether $\mathrm{Epi}_{L(\chi)}(\Gamma(\chi), G)$ is nonempty, and if yes whether χ itself is of the form $\mathrm{Tr}(\Phi)$ and not only $\chi + \overline{\chi} = \mathrm{Tr}(\Phi) + \overline{\mathrm{Tr}(\Phi)}$ holds.

23.1. Classification for Large $[\chi, 1_G]$. A necessary condition on the existence of $\Phi \in \mathrm{Hom}_{L(\chi)}(\Gamma(\chi), G)$ is that $\prod_{i=1}^{r} \Phi(c_i)$ is contained in G'. The following lemma shows that the condition (RH) implies this property in the case that G/G' is an elementary abelian 2-group.

LEMMA 23.1. *Let G be a group, $N \leq G$ a normal subgroup such that G/N is an elementary abelian 2-group, χ an (RH)-character of G, $L(\chi) = (L_1, L_2, \dots, L_r)$, and $\sigma_i \in L_i$ for $1 \leq i \leq r$. Then $\prod_{i=1}^{r} \sigma_i \in N$.*

Proof. For each nontrivial linear character φ of G that contains N in its kernel, the proof of Lemma 22.16 and Example 22.20 yield that

$$
\begin{aligned}
2 \cdot [\chi, \varphi] &= v_\varphi + \sum_{H \in \overline{CY}(G)/\sim_G} l_\chi(H) \cdot T_{H,\varphi} \\
&= -2 + 2 \cdot l_\chi(1) + \sum_{\substack{H \in CY(G)/\sim_G \\ H \not\subseteq \ker(\varphi)}} l_\chi(H),
\end{aligned}
$$

so the sum on the right hand side is even. By the choice of the σ_i, it is equal to the number of σ_i outside $\ker(\varphi)$. This implies that $\prod_{i=1}^{r} \sigma_i$ lies in the intersection of these kernels, which is N (cf. [**Isa76**, p. 23 and Cor. 2.23]). $\square$

Applied to the case $N = G'$, the lemma guarantees that $\mathrm{Hom}_{L(\chi)}(\Gamma(\chi), G)$ (and also $\mathrm{Epi}_{L(\chi)}(\Gamma(\chi), G)$) is nonempty if G is an ambivalent group and $[\chi, 1_G]$ is large enough. In order to get a quantitative formulation, we define

$G^{[i]} = \{\sigma \in G' \mid \sigma \text{ is a product of at most } i \text{ commutators of elements in } G\}$

and $c.l.(G) = \min\{i \in \mathbb{Z} \mid G^{[i]} = G'\}$, which stands for "commutator length". Note that $G^{[i]}$ is in general *not* a group.

THEOREM 23.2. *Let G be a finite group whose commutator factor group is an elementary abelian 2-group, and assume that G can be generated by n elements. Let χ be an (RH)-character of G such that $[\chi, 1_G] \geq c.l.(G) + n/2$. Then there is a $\Phi \in \mathrm{Epi}_{L(\chi)}(\Gamma(\chi), G)$ such that $\mathrm{Tr}(\Phi) + \overline{\mathrm{Tr}(\Phi)} = \chi + \overline{\chi}$.*

Proof. Let $m = 2\lfloor(n+1)/2\rfloor$, the smallest even number that is larger than or equal to n, and choose m elements $\tau_1, \tau_2, \ldots, \tau_m$ that generate G, and choose $r = \sum_{H \in CY(G)/\sim_G} l_\chi(H)$ elements $\sigma_1, \sigma_2, \ldots, \sigma_r$ in G such that $l_\chi(H)$ of the $\langle \sigma_i \rangle$ are G-conjugate to H, for all $H \in CY(G)$. Then $\prod_{i=1}^r \sigma_i \in G'$ by Lemma 23.1. Set $g = [\chi, 1_G] - m/2$. Because $g \geq c.l.(G)$, there are elements $\alpha_1, \alpha_2, \ldots, \alpha_g, \beta_1, \beta_2, \ldots, \beta_g$ in G such that

$$\left([\tau_1, \tau_2] \cdots [\tau_{m-1}, \tau_m]\sigma_1 \cdots \sigma_r\right)^{-1} = [\alpha_1, \beta_1] \cdots [\alpha_g, \beta_g].$$

Let $\Phi \colon \Gamma(m/2 + g; |\sigma_1|, |\sigma_2|, \ldots, |\sigma_r|) \to G$ be the surface kernel epimorphism given by the images $\alpha_i, \beta_i, \tau_j, \sigma_k$, with $1 \leq i \leq g$, $1 \leq j \leq m$, $1 \leq k \leq r$. Then $\mathrm{Tr}(\Phi) + \overline{\mathrm{Tr}(\Phi)} = \chi + \overline{\chi}$.   □

COROLLARY 23.3. *If G is an ambivalent group that can be generated by n elements then a G-character χ with $[\chi, 1_G] \geq c.l.(G) + n/2$ comes from a Riemann surface if and only if it satisfies the condition* (RH).

REMARK 23.4. All $G^{[i]}$ and hence $c.l.(G)$ can be computed from the character table of G. That is, $G^{[1]}$ is the set of commutators in G, and an element $\sigma \in G$ is in this set if and only if

$$\sum_{\chi \in \mathrm{Irr}(G)} \frac{\chi(\sigma)}{\chi(1)} \neq 0$$

by [**Isa76**, Problem (3.10)]. For $i > 1$, we have

$$G^{[i]} = \{\sigma\tau \mid \sigma \in G^{[i-1]}, \tau \in G^{[1]}\},$$

so $\sigma \in G^{[i]}$ if and only if $\mathrm{Hom}_L(0, G) \neq \emptyset$, where $L = (G^{[i-1]}, G^{[1]}, (\sigma^{-1})^G)$, and the cardinality of $\mathrm{Hom}_L(0, G)$ can be computed from the character table of G by Theorem 14.1.

The number $c.l.(G)$ can be arbitrarily large. For example [**Isa77**] shows a class of perfect groups with unbounded commutator length. For many finite simple groups G it is known that $c.l.(G) = 1$, i.e., every element in G' is a commutator of elements in G. It is conjectured that this holds for *all* finite simple groups, see for example [**Bon93**].

23.2. Infinite Numbers of Exceptions. If $[\chi, 1_G]$ is small then the condition (RH) is in general not sufficient for a real character χ to come from a Riemann surface, as we have seen in Example 22.12. Moreover, in general we cannot expect that for a given group G, only *finitely* many G-characters satisfy the condition (RH) but do not come from a Riemann surface. In this section, we describe two situations where infinitely many (RH)-characters of a given group do not come from Riemann surfaces.

THEOREM 23.5. *Let G be either a nonabelian group with nontrivial center or an abelian group that cannot be generated by three elements, and Z a central subgroup in G of prime order p.*

Then each character $\chi = 1_G + n \cdot (\rho_Z - 1_Z)^G$, for any positive integer n, satisfies the condition (RH) but does not come from a Riemann surface.

Proof. By Corollary 22.5, χ satisfies the condition (RH), with $g_0 = 1$, $l_\chi(Z) = 2n$, and $l_\chi(H) = 0$ for all $H \in CY(G) \setminus \{Z\}$.

Suppose that $\chi = \mathrm{Tr}(\Phi)$ for $\Phi \in \mathrm{Epi}_{L(\chi)}(\Gamma(\chi), G)$. Then

$$\Gamma(\chi) = \Gamma(1; p, p, \ldots, p) = \langle a, b, c_1, c_2, \ldots, c_{2n} \rangle,$$

where a and b are hyperbolic generators and the c_i are elliptic generators of order p. We have

$$[\Phi(a), \Phi(b)] \cdot \prod_{i=1}^{2n} \Phi(c_i) = 1,$$

and we want to show that already the product of the $\Phi(c_i)$ is the identity. For $u \in I(p)$, let u' denote the unique element of $I(p)$ for which $u \cdot u' \equiv 1$ (mod p). The $\Phi(c_i)$ are nonidentity powers of z, with $Z = \langle z \rangle$, and we have

$$\{i \mid 1 \le i \le 2n, \Phi(c_i) = z^{u'}\} = \{i \mid 1 \le i \le 2n, \Phi(c_i)^u = z\} = \frac{|\mathrm{Fix}^G_{X,u}(z)|}{[C_G(z):Z]}$$

by Lemma 11.5, where $X = \mathcal{U}/\ker(\Phi)$ is the Riemann surface associated with the character χ. Now

$$\chi(z) = 1 + \sum_{u=1}^{p-1} |\mathrm{Fix}_{X,u}(z)| \cdot \frac{\zeta_p^u}{1 - \zeta_p^u}$$

by Theorem 12.1, and we can write the expression on the right hand side as

$$\left(1 - \sum_{u=1}^{(p-1)/2} |\mathrm{Fix}_{X,p-u}(z)|\right) + \sum_{u=1}^{(p-1)/2} (|\mathrm{Fix}_{X,u}(z)| - |\mathrm{Fix}_{X,p-u}(z)|) \cdot \frac{\zeta_p^u}{1 - \zeta_p^u}$$

by Lemma 12.2. Since $\chi(z)$ is a rational integer, Theorem C.2 yields that $|\mathrm{Fix}_{X,u}(z)| = |\mathrm{Fix}_{X,p-u}(z)|$ for all $1 \le u \le (p-1)/2$.

Every element in $G \setminus Z$ acts fixed point freely on X, so we have $\mathrm{Fix}_{X,u}(z) = \mathrm{Fix}^G_{X,u}(z)$ for all $u \in I(p)$. If we set $f(u) = p|\mathrm{Fix}_{X,u}(z)|/|G|$ then we get

$$\prod_{i=1}^{2n} \Phi(c_i) = \prod_{u'=1}^{p-1} \left(z^{u'}\right)^{f(u)} = \prod_{u'=1}^{(p-1)/2} \left(z^{u'f(u)} \cdot z^{(p-u)'f(p-u)}\right),$$

and each factor of the last product is the identity because all $f(u)$ are integers, $f(u) = f(p - u)$, and $(p - u)' = p - u'$.

This means that both $\prod_{i=1}^{2n} \Phi(c_i)$ and $[\Phi(a), \Phi(b)]$ are the identity, hence $\Phi(a)$ and $\Phi(b)$ commute, and $G = \Phi(\Gamma(\chi)) = \langle \Phi(a), \Phi(b), z \rangle$ is an abelian group that can be generated by three elements, contrary to our assumptions. $\square$

EXAMPLE 23.6 (cf. [**Kim93**]). The smallest groups for which Theorem 23.5 applies are the two nonabelian groups of order 8, the dihedral group $D_8 = \langle a, b \mid b^{-1}aba, a^4, b^2 \rangle$ and the quaternion group $Q_8 = \langle a, b \mid b^{-1}aba, a^4, a^2b^2 \rangle$. Let G be one of these groups, and in either case, let $z = a^2$ denote the unique central involution. Both groups have the same irreducible characters, they are shown in Table 15.

	1	z	a	b	ab
1_G	1	1	1	1	1
χ_2	1	1	1	-1	-1
χ_3	1	1	-1	1	-1
χ_4	1	1	-1	-1	1
χ_5	2	-2	0	0	0

TABLE 15. Irreducible Characters of D_8 and Q_8

The (RH)-characters of G that do not come from Riemann surfaces, according to Theorem 23.5, are of the form

$$1_G + n \cdot (\rho_{\langle z \rangle} - 1_{\langle z \rangle})^G = 1_G + 2n \cdot \chi_5.$$

The central subgroup $\langle z \rangle$ is of order 2, and for this special case, the proof of Theorem 23.5 is particularly easy because $\Phi(c_i) = z$ for all $2n$ elliptic generators c_i, so we do not need Theorem C.2 to conclude that $\prod_{i=1}^{2n} \Phi(c_i) = 1$ holds.

For a further inspection of the groups D_8 and Q_8, see Section 26.   ◇

In the spirit of this example, we can generalize the theorem as follows.

THEOREM 23.7. *Let G be a finite group, N a normal subgroup of G, and $\pi\colon G \to F$ the natural epimorphism. Assume that F is either a nonabelian group with nontrivial center of even order or an abelian group of even order that cannot be generated by three elements. Let Z be a central subgroup in F of order 2, and $U \le G$ the preimage of Z under π.*

Then each (RH)-character χ of G with the properties that $[\chi, 1_G] = 1$, $l_\chi(H) = 0$ for $H \nleq U$, and

$$\sum_{H \subseteq U, H \nleq N} l_\chi(H) \equiv 0 \pmod 2,$$

does not come from a Riemann surface.

Proof. Suppose that $\chi = \mathrm{Tr}(\Phi)$ for a surface kernel epimorphism $\Phi\colon \Gamma \to G$, and let a and b denote the images of the hyperbolic generators of Γ. Consider the composition $\overline{\Phi} = \pi \circ \Phi$. Each elliptic generator of Γ is mapped into

$Z = U/N$ under $\overline{\Phi}$, and since Z is of order 2 and the number of nonidentity images is even, their product is the identity in F. So $\overline{\Phi}(a)$ and $\overline{\Phi}(b)$ commute, hence $\langle \overline{\Phi}(a), \overline{\Phi}(b), Z \rangle$ is abelian and generated by three elements. By our assumptions on F, this means that $\overline{\Phi}$ cannot be surjective.   □

THEOREM 23.8. *Let G be a finite nonabelian group, N a proper normal subgroup of G', and χ an (RH)-character of G such that $[\chi, 1_G] = 1$ and l_χ vanishes outside N. Then χ does not come from a Riemann surface.*

Proof. Suppose that $\chi = \text{Tr}(\Phi)$ for a surface kernel epimorphism $\Phi \colon \Gamma \to G$, and let a and b denote the images of the hyperbolic generators of Γ. The images of all elliptic generators of Γ lie in N, so the commutator $[a, b]$ also does. $\Phi(\Gamma)$ is contained in $U = \langle N, a, b \rangle$, and U' can be obtained as the normal closure of the commutators of group generators of U. But $[N, a]$, $[N, b]$, $[N, N]$, and $[a, b]$ are all contained in N, so $U' \subseteq N$. This means in particular that $U \neq G$, hence Φ cannot be surjective.   □

23.3. Finite Numbers of Exceptions. We have seen examples of groups G for which all (RH)-characters come from Riemann surfaces, and we have seen examples where infinitely many (RH)-characters fail to come from Riemann surfaces. Now we want to show groups for which all except at most finitely many (RH)-characters come from Riemann surfaces. The situations of Section 23.2 involve certain normal subgroups of the group in question, a central subgroup in Theorem 23.5 and a nontrivial proper normal subgroup inside the derived subgroup in Theorem 23.8. Trying to avoid such normal subgroups, we look at simple groups.

The finiteness property we will present is based on the possibility of writing each $\sigma \in G$ as a product of a certain number of elements in the rational conjugacy class of σ, which is defined as $\{\tau \in G \mid \langle \sigma \rangle \sim_G \langle \tau \rangle\}$. Let $rep(H^G)$ denote the smallest integer such that for all $m \geq rep(H^G)$, a generator σ of H can be written as a product of exactly m elements in the rational class of σ; if no such number exists then we set $rep(H^G) = \infty$.

REMARK 23.9. If σ can be written as a product of m elements in the rational class of σ then it can be written as a product of $m + 2$ elements, because we may multiply by $\sigma \cdot \sigma^{-1}$. So we can write each σ as a product of arbitrary odd numbers of elements in the rational class of σ. If σ has *odd* order m then $\sigma = \sigma^{m+1}$ shows that $rep(\langle \sigma \rangle^G) \leq m$.

DEFINITION 23.10. For a given group G and $\sigma \in G$, define $n(\sigma)$ as the smallest integer such that each element in $\langle G', \sigma \rangle$ can be written as a product of at most $n(\sigma)$ elements in the rational class C of σ in G. If no such integer exists then set $n(\sigma) = \infty$.

Let $k(\sigma)$ denote the smallest integer such that $\langle G', \sigma \rangle$ can be generated by $k(\sigma)$ elements in C. If no such integer exists then set $k(\sigma) = \infty$.

If G is a simple group and $\sigma \in G^\times$, no proper subgroup of G contains the whole conjugacy class σ^G. This is clear from the fact that the group generated by σ^G is closed under G-conjugation, and hence is a normal subgroup of G. So $n(\sigma)$ and $k(\sigma)$ are finite for all $\sigma \in G^\times$.

More generally, if G is a group such that G' is simple and contained in every nontrivial proper normal subgroup then the conjugacy class σ^G generates a group containing $\langle G', \sigma \rangle$, hence $n(\sigma)$ and $k(\sigma)$ are finite in this case also.

REMARK 23.11. We can write σ as a product of n elements in its rational class C if and only if $\mathrm{Hom}_L(0, G) \neq \emptyset$, where $L = (C, C, \ldots, C)$, with $n + 1$ occurrences of C. So $rep(\langle \sigma \rangle^G)$ can be computed from the character table of G, using Theorem 14.1. Analogously, the character table of G suffices to compute $n(\sigma)$.

But the character table does *not* in general provide enough information to compute $k(\sigma)$. This is because the computation of $k(\sigma)$ involves a question of generation, which can be translated only into the question whether a certain set $\mathrm{Epi}_L(0, G)$ is nonempty, and this is in general not determined by the character table of G.

EXAMPLE 23.12. Table 16 lists the values of $rep_{\max} = \max\{rep(H^G) \mid H \in CY(G)\}$ and $n_{\max} = \max\{n(\sigma) \mid \sigma \in G^\times\}$ for the sporadic simple groups G. The values were computed using the character tables in GAP.   ◇

G	$rep_{\max}$	$n_{\max}$	G	$rep_{\max}$	$n_{\max}$	G	$rep_{\max}$	$n_{\max}$
M_{11}	1	3	McL	1	3	Ly	1	3
M_{12}	1	4	He	1	4	Th	1	3
J_1	1	2	Ru	1	3	Fi_{23}	5	6
M_{22}	1	3	Suz	1	4	Co_1	1	4
J_2	1	4	ON	1	3	J_4	1	3
M_{23}	1	3	Co_3	3	3	F_{3+}	1	3
HS	3	3	Co_2	3	4	B	3	4
J_3	1	3	Fi_{22}	5	6	M	1	3
M_{24}	1	3	HN	1	4			

TABLE 16. Values of $rep_{\max}$ and $n_{\max}$ for Sporadic Simple Groups

THEOREM 23.13. *Let G be a finite simple group, and assume that $rep(H^G) < \infty$ for all $H \in CY(G)$. Then for all except at most finitely many (RH)-characters χ of G, there is $\Phi \in \mathrm{Epi}_{L(\chi)}(\Gamma(\chi), G)$ such that $\Phi + \overline{\Phi} = \chi + \overline{\chi}$.*

If additionally G is ambivalent then at most finitely many (RH)-characters of G do not come from Riemann surfaces.

Proof. Let χ be an (RH)-character of G such that $l_\chi(\sigma) \geq n(\sigma) + k(\sigma) + rep(\langle\sigma\rangle^G) - 1$ holds for an element $\sigma \in G$. We claim that there is a $\Phi \in \mathrm{Epi}_{L(\chi)}(\Gamma(\chi), G)$ such that $\chi + \overline{\chi} = \mathrm{Tr}(\Phi) + \overline{\mathrm{Tr}(\Phi)}$.

To define Φ, we choose the identity of G as image of the hyperbolic generators of $\Gamma(\chi)$. Then we choose, for $K \in CY(G)/\sim_G$ with $K \not\sim_G \langle\sigma\rangle$, arbitrary images $\tau_{i,K}$, $1 \leq i \leq l_\chi(K)$, among the G-conjugates of generators of K. Then we set

$$\tau = \prod_{\substack{K \in CY(G)/\sim_G \\ K \not\sim_G \langle\sigma\rangle}} \prod_{i=1}^{l_\chi(K)} \tau_{i,K},$$

and choose $k = k(\sigma)$ images σ_1, σ_2, ... , σ_k in the rational class C of σ that generate G. Then $(\tau\sigma_1\sigma_2\cdots\sigma_k)^{-1}$ is a product of $m \leq n(\sigma)$ elements σ_{k+1}, σ_{k+2}, ... , σ_{k+m} in σ^G, so $\tau\sigma_1\sigma_2\cdots\sigma_{k+m}$ is the identity.

There are still $l = l_\chi(\sigma) - k - m$ images in C to be defined, and because $l \geq rep(\langle\sigma\rangle^G) - 1$, we can replace σ_1 by $l + 1$ elements in its rational class. Choosing the images $\Phi(c_i)$ this way, we get $\Phi \in \mathrm{Epi}_{L(\chi)}(\Gamma(\chi), G)$.

Now the statement of the theorem follows from the facts that only finitely many values of $g_0 = [\chi, 1_G]$ must be considered by Theorem 23.2, that for each of them, at most finitely many vectors $(l_\chi(K))_{K \in CY(G)/\sim_G}$ do not admit the construction of Φ as above, and for each such vector, there are at most finitely many G-characters ψ with $\psi + \overline{\psi} = \chi + \overline{\chi}$. □

REMARK 23.14. Note that Theorem 23.13 holds not only for *nonabelian* simple groups but also for cyclic groups of odd prime order.

Now we generalize the idea of Theorem 23.13 to certain situations where G is a group such that G' is simple and G/G' is an elementary abelian 2-group. Clearly $rep(\langle\sigma\rangle^G) = \infty$ for all $\sigma \in G \setminus G'$ in such a case, because $\sigma = \sigma_1\sigma_2\cdots\sigma_n$ for $\langle\sigma_i\rangle \sim_G \langle\sigma\rangle$ implies $\sigma G' = (\sigma_1 G')(\sigma_2 G')\cdots(\sigma_n G') = (\sigma G')^n$, hence n is necessarily odd. But we need the finiteness condition on $rep(H^G)$ only for subgroups H of G'.

THEOREM 23.15. *Let G be a finite group such that G' is a simple group that is contained in every nontrivial proper normal subgroup of G, and such that G/G' is an elementary abelian 2-group. Assume that $rep(H^{G'}) < \infty$ for all $H \in CY(G)$ with $H \leq G'$.*

Then for all except at most finitely many (RH)-characters χ of G with the property that $g_0 = [\chi, 1_G]$ is either zero or satisfies $2^{2g_0} \geq [G{:}G']$, there is $\Phi \in \mathrm{Epi}_{L(\chi)}(\Gamma(\chi), G)$ such that $\Phi + \overline{\Phi} = \chi + \overline{\chi}$.

If additionally G is ambivalent and $[G{:}G'] \leq 4$ then at most finitely many (RH)-characters of G do not come from Riemann surfaces.

Proof. First let χ be an (RH)-character of G such that $l_\chi(\sigma) \geq n(\sigma) + k(\sigma) + rep(\langle\sigma\rangle^G) - 1$ for an element $\sigma \in G'$. Then we can construct the homomorphism Φ as in the proof of Theorem 23.13. Note that the product of the images of the elliptic generators lies in G' by Lemma 23.1. The surjectivity of Φ for $g_0 = 0$ follows from Lemma 22.19. For $2^{2g_0} \geq [G{:}G']$, we may choose images of the $2g_0$ hyperbolic generators that generate G modulo G'.

Now let χ be an (RH)-character of G such that $l_\chi(\sigma) \geq n(\sigma) + k(\sigma) + rep(\langle\sigma\rangle^G) - 1$ for an element $\sigma \in G \setminus G'$. Similarly to the proof of Theorem 23.13 and of the case above, we define a homomorphism $\Phi\colon \Gamma(\chi) \to G$. We choose arbitrary images in the appropriate rational classes for those elliptic generators of $\Gamma(\chi)$ that are mapped to generators of cyclic groups $K \not\sim_G \langle\sigma\rangle$; let τ denote the product of these elements. By Lemma 23.1, we have $\tau \in \langle G', \sigma\rangle$. Next we choose $k = k(\sigma)$ images $\sigma_1, \sigma_2, \ldots, \sigma_k$ in the rational class of σ that generate $\langle G', \sigma\rangle$. Then $(\tau\sigma_1\sigma_2\cdots\sigma_k)^{-1}$ lies in $\langle G', \sigma\rangle$, so is a product of $m \leq n(\sigma)$ elements $\sigma_{k+1}, \sigma_{k+2}, \ldots, \sigma_{k+m}$ in C. Hence $\tau\sigma_1\sigma_2\cdots\sigma_{k+m}$ is the identity. Now $l = l_\chi(\sigma) - k - m$ images in C are left to be defined. Because of Lemma 23.1, l is even, so we may choose $l/2$ images σ and σ^{-1} each. Thus we get a homomorphism Φ, which is surjective for $g_0 = 0$ or $2g_0 \geq [G{:}G']$ by Lemma 22.19 or by an appropriate choice of images of the hyperbolic generators of $\Gamma(\chi)$, respectively.

Now the statement of the theorem follows as in the proof of Theorem 23.13. $\square$

24. A Classification Algorithm

We describe an algorithm to classify, for a given ambivalent group G, those (RH)-characters of G that do not come from Riemann surfaces.

First we note that the infinite series of exceptions described in Theorems 23.5, 23.7, and 23.8 can be computed from the character table of G; we just have to calculate the unions of conjugacy classes forming those normal subgroups N that are central of prime order (in an appropriate factor group of G) or maximal with the property of being contained properly in the derived subgroup of G. If we set $E(N) = \{H \in CY(G) \mid H \not\subseteq N\}$ then the series of exceptions arising from N, according to Theorem 23.5 or Theorem 23.8, is described by the system of equations $l_\chi(H) = 0$ for $H \in E(N)$; Theorem 23.7 involves two normal subgroups U and N with U/N of order 2, and the corresponding infinite series of exceptions is given by the equations $l_\chi(H) = 0$ for $H \in E(U)$ and the congruence $\sum_{H \subseteq U, H \not\subseteq N} l_\chi(H) \equiv 0 \pmod{2}$.

Now we consider the problem of computing those (RH)-characters that do not lie in the above infinite series but also do not come from Riemann surfaces. Recall the equation

$$(a_\varphi)_{\varphi \in \mathrm{Rat}(G)} = v + (l_\chi(H))_{H \in \widehat{CY}(G)/\sim_G} \cdot T$$

from Lemma 22.16, for any class function χ of G that satisfies condition (E), where

$$2\chi = \chi + \overline{\chi} = 2 \cdot 1_G + 2(g_0 - 1) \cdot \rho_G + \sum_{H \in CY(G)/\sim_G} l_\chi(H) \cdot (\rho_G - 1_H^G)$$

and $g_0 = l_\chi(H_0)$, with H_0 the trivial subgroup of G.

If χ is an (RH)-character then clearly all $l_\chi(H)$ are nonnegative integers, and all a_φ are nonnegative and even. For the converse, we take the same viewpoint as in Example 22.21. We will refer to this observation several times, so we formulate it as a lemma.

LEMMA 24.1. *Let $l = (l(H))_{H \in \widehat{CY}(G)/\sim_G}$ be a vector of nonnegative integers. If the corresponding class function $\chi = \frac{1}{2} \sum_{\varphi \in \mathrm{Rat}(G)} a_\varphi \cdot \varphi$ with $\mathrm{Rat}(G)$-coefficients $a = v + l \cdot T$ is not an (RH)-character then at least one of the following cases occurs:*

(i) *a has an odd entry;*
(ii) *a has a negative entry;*
(iii) *$\chi(1) < 2$.*

These conditions can be formulated as congruences and inequalities in terms of the $l(H)$. The inequalities are trivial if $g_0 \geq 2$, and they exclude exactly the vector with $l(H) = 0$ for all $H \in CY(G)$ if $g_0 = 1$.

The following rather technical lemmas provide the tools for the algorithm we are interested in. Our problem is to decide when we have found all exceptional characters. In the situations of Theorems 23.13 and 23.15, we could use the bounds derived in the proofs, but we are interested also in groups that admit infinitely many exceptions, as described in Section 23.2.

First we deal with those vectors l for which $l(H)$ is sufficiently large for a cyclic subgroup H that does not lie in a proper normal subgroup of G. This suffices already to treat simple groups. We will see that, compared with the general case, only relatively few tests are necessary, due to the fact that the interesting values $l(K)$ for $K \not\sim_G H$ are a priori bounded. Note that a nonzero value $l(H)$ automatically excludes the infinite series of exceptions mentioned above. We call l an (RH)-vector if its corresponding class function is an (RH)-character.

LEMMA 24.2. *Let G be an ambivalent group, g_0 a nonnegative integer, and $H \in CY(G)$ not contained in a proper normal subgroup of G. For a nonnegative integer q, we define $M(q, H)$ as the set of those vectors $(l(K))_{K \in \widehat{CY}(G)/\sim_G}$ with the properties $l(H_0) = g_0$, $l(H) \in \{q, q + 1\}$, and $l(K) \in \{0, 1\}$ for $K \not\sim_G H$.*

Suppose that each character corresponding to an (RH)-vector in $M(q, H)$ comes from a Riemann surface, and that no vector in $M(q, H)$ fails to be an (RH)-vector because of the condition 24.1 (iii). If $g_0 = 0$ then additionally assume that

$$q \geq \max \left\{ \frac{-v_\varphi}{T_{H,\varphi}} \bigg| \varphi \in \text{Rat}(G), \varphi \neq 1_G \right\}.$$

(Note that for the chosen H, $T_{H,\varphi} \neq 0$ for each nontrivial $\varphi \in \text{Rat}(G)$.)

Then each character corresponding to an (RH)-vector l with $l(H) \geq q$ comes from a Riemann surface.

Proof. Let l be an (RH)-vector with $l(H) \geq q$, and consider the (unique) vector $l' \in M(q, H)$ that is given by the properties $l'(K) \equiv l(K) \pmod{2}$ for all $K \in \widehat{CY}(G)/ \sim_G$.

We claim that l' is an (RH)-vector. To show this, we exclude the possibilities 24.1 (i)–(iii). All entries in $v + l \cdot T$ and $l - l'$ are even because l is an (RH)-vector by assumption and $l' \equiv l \pmod{2}$ by construction, so $v + l' \cdot T = v + l \cdot T - (l - l') \cdot T$ has only even entries, and hence case (i) cannot occur. Case (ii) is possible only for $g_0 = 0$, but then the additional assumption on q yields, for any nontrivial $\varphi \in \text{Rat}(G)$, that

$$
\begin{aligned}
v_\varphi + (l' \cdot T)_\varphi &= v_\varphi + \sum_{K \in CY(G)/\sim_G} l'(K) \cdot T_{K,\varphi} \\
&\geq v_\varphi + l'(H) \cdot T_{H,\varphi} \\
&\geq v_\varphi + q \cdot T_{H,\varphi} \\
&\geq v_\varphi - v_\varphi = 0
\end{aligned}
$$

holds. Finally, case (iii) cannot occur because this is explicitly assumed.

So the character corresponding to l' comes from a Riemann surface, and we can construct a surface-kernel epimorphism for the character corresponding to l from one for l' by adding appropriate pairs of mutually inverse elements as images of the additional elliptic generators. $\square$

Now we look at the general case, where the subgroup H in question may be contained in a proper normal subgroup. Again, the idea is to take an (RH)-vector l with sufficiently large $l(H)$, and to find a smaller (RH)-vector l' for which a Riemann surface has already been established. But here we cannot in general reduce all entries $l(K)$ for $K \not\sim_G H$. We proceed iteratively, by considering vectors $l = (l(H))_{H \in \widehat{CY}(G)/\sim_G}$ for increasing values of $\sum_{H \in CY(G)/\sim_G} l(H)$, which we call the *norm* of l.

LEMMA 24.3. *Let G be an ambivalent group, and g_0 and N nonnegative integers. If $g_0 = 0$ then, for $K \in CY(G)$, set*

$$q(K) = \max\left\{\frac{-v_\varphi}{T_{K,\varphi}}\middle|\varphi \in \mathrm{Rat}(G), T_{K,\varphi} \neq 0\right\},$$

otherwise set $q(K) = 1$. Suppose that for each (RH)-vector of norm equal to N and $N + 1$, the corresponding character either comes from a Riemann surface or lies in one of the infinite series of exceptions mentioned above. Additionally suppose that no vector of norm N or $N + 1$ fails to be an (RH)-vector because of 24.1 (iii). Then each character corresponding to an (RH)-vector l of norm at least N comes from a Riemann surface or lies in one of the above infinite series of exceptions, provided that $l(H) \geq q(H) + 2$ for at least one $H \in CY(G)$.

Proof. Let $l = (l(K))_{K \in \widehat{CY}(G)/\sim_G}$ be an (RH)-vector of norm at least $N+2$ that does not lie in an infinite series of exceptions, and such that $l(H) \geq q(H) + 2$. We have to show that the character corresponding to l comes from a Riemann surface.

Consider the vector l' that is obtained from l by decreasing $l(H)$ by 2. If l' is an (RH)-vector then we are done by induction on N, since we can construct a surface kernel epimorphism for l from one for l', in the same way as in the proof of Lemma 24.2. Note that l' does not belong to an infinite series of exceptions since l and l' vanish on the same subset of $CY(G)/\sim_G$, and the sums of a subset of values of l and l' have the same parity.

To see that l' is an (RH)-vector, we exclude the cases 24.1 (i)–(iii) in the same way as in the proof of Lemma 24.2. Again, case (iii) is explicitly excluded, and case (i) cannot occur because l is an (RH)-vector and $v + l \cdot T$ differs from $v + l' \cdot T$ by $(2 \cdot T_{H,\varphi})_{\varphi \in \mathrm{Rat}(G)}$, which has only even entries. For case (ii), we have to take care of the possibility that $T_{H,\varphi} = 0$ for a nontrivial $\varphi \in \mathrm{Rat}(G)$; but then

$$(v + l' \cdot T)_\varphi = (v + l \cdot T)_\varphi \geq 0$$

holds. The proof for $T_{H,\varphi} \neq 0$ is the same as in Lemma 24.2, we get

$$v_\varphi + (l' \cdot T)_\varphi \geq v_\varphi + l'(H) \cdot T_{H,\varphi} \geq v_\varphi + q(H) \cdot T_{H,\varphi} \geq 0,$$

by the choice of the values $q(K)$. □

Now we are ready to formulate an algorithm.

ALGORITHM 24.4. Let G be an ambivalent group, and g_0 a nonnegative integer. If at most finitely many (RH)-vectors $l = (l(H))_{H \in \widehat{CY}(G)/\sim_G}$ of G with $l(H_0) = g_0$ do not come from Riemann surfaces and do not lie in the infinite series of exceptions described in Theorems 23.5, 23.7, and 23.8 then this algorithm computes the set V of these vectors.

1. If $g_0 = 1$ then compute the vectors l corresponding to the infinite series of exceptions, and set I to be the set of these vectors. For $g_0 \neq 1$, set $I = \emptyset$.
2. Compute the numbers $q(H)$ defined in Lemma 24.3, for $H \in CY(G)/\sim_G$. Set $N = \min\{q(H) \mid H \in CY(G)/\sim_G\}$, $V = \emptyset$, and $S = \emptyset$.
3. Loop over all those vectors l with $0 \leq l(H) \leq q(H) + 1$ for all $H \in CY(G)/\sim_G$. Add those l to S that fail to be (RH)-vectors because of 24.1 (iii). Add those l to V that are (RH)-vectors, not contained in I, and such that the corresponding character does not come from a Riemann surface.
4. If $V \cup S$ contains no vector of norm N or $N+1$ then return V and terminate the algorithm.
5. Compute the set $E \subseteq CY(G)/\sim_G$ of those groups that are not contained in any proper normal subgroup of G. For each $H \in E$, compute the smallest value $b(H) \geq q(H)$ such that $M(b(H), H)$ satisfies the assumptions of Lemma 24.2.
6. Increase N by 1. Loop over all those vectors l of norm $N+1$ with $l(H) > q(H) + 1$ for at least one $H \in CY(G)/\sim_G$ and with $l(H) < b(H)$ for all $H \in E$. Add those l to S that fail to be (RH)-vectors because of 24.1 (iii). Add those l to V that are (RH)-vectors, not contained in I, and such that the corresponding character does not come from a Riemann surface.
7. If $V \cup S$ contains no vector of norm N or $N+1$ then return V and terminate the algorithm. Otherwise go to Step 6.

The validity of the algorithm can be shown as follows. Suppose the algorithm terminates with value N, and suppose that l is one of the finitely many (RH)-vectors not contained in I whose characters do not come from a Riemann surface. We have to show that l lies in V. If the algorithm was terminated in Step 4 then all vectors of norm up to $N+1$ have been inspected in Step 3, and Lemma 24.3 guarantees that l has norm smaller than N or $l(H) \leq q(H) + 1$ for all $H \in CY(G)/\sim_G$. Because $N \leq q(H)$ for all H, l is one of the vectors checked in Step 3. If the algorithm was terminated in Step 7 then all exceptional vectors of norms N and $N+1$ are contained in $V \cup S$, because each vector of norm at most $N+1$ either has been checked in Step 3 or 6 or is established by Lemma 24.2; thus again we apply Lemma 24.3, and get that

l has norm smaller than N or $l(H) \leq q(H) + 1$ for all $H \in CY(G)/\sim_G$. But all exceptional vectors with these properties have been found in Step 3 or 6.

REMARK 24.5. In order to make the checks in Step 3 and 6 as cheap as possible, we consider a partial ordering $\prec$ on the vectors l, with the property that if the character of l' comes from a Riemann surface and $l' \prec l$ then l also comes from a Riemann surface. By the argument used in the proofs of Lemma 24.2 and Lemma 24.3, we may define this partial ordering by $l' \prec l$ if $l'(H) \leq l(H)$ and $l'(H) \equiv l(H) \pmod 2$ holds for all $H \in CY(G)/\sim_G$. For the case that the rational class C containing the generators of H satisfies $\mathrm{Hom}_{(C,C,C)}(0, G) \neq \emptyset$, the requirement $l'(H) \equiv l(H) \pmod 2$ may be weakened to $l'(H) \equiv l(H) \pmod 2$ or $l'(H) > 0$, since in the latter case we may replace one element in C by a product of two elements in C.

Thus we can compute in Algorithm 24.4 a list of those minimal vectors w.r.t. the partial ordering $\prec$ whose characters come from Riemann surfaces. If the algorithm terminates then this list of representatives and the list of infinite series can be returned together with the set V of exceptions. This then provides a proof of the classification for G.

EXAMPLE 24.6. We classify the characters of the symmetric group $G = S_4$ on four points. Table 17 shows the character table and the matrix T of G, with rows and columns arranged compatibly; the classes of G are denoted by names composed from the element order and a distinguishing letter, so 2A is the class of double transpositions, and 2B is the class of transpositions.

	1A	2A	3A	2B	4A
1_G	1	1	1	1	1
χ_2	1	1	1	-1	-1
χ_3	2	2	-1	0	0
χ_4	3	-1	0	1	-1
χ_5	3	-1	0	-1	1

$$\begin{pmatrix} 2 & 2 & 4 & 6 & 6 \\ 0 & 0 & 0 & 2 & 2 \\ 0 & 0 & 2 & 2 & 2 \\ 0 & 1 & 1 & 1 & 2 \\ 0 & 1 & 1 & 3 & 2 \end{pmatrix}$$

TABLE 17. Character Table and Matrix T for S_4

There is exactly one infinite series of exceptions, according to Theorem 23.8, which is given by $l_\chi(3A) = l_\chi(2B) = l_\chi(4A) = 0$, the characters are of the form $\chi = 1_G + n(\chi_4 + \chi_5)$.

First we consider the case $g_0 = 0$. The values $q(H)$ are $q(2A) = q(3A) = 3$, $q(2B) = 6$, and $q(4A) = 4$. The condition $\mathrm{Hom}_{(C,C,C)}(0, G) \neq \emptyset$ is satisfied for the classes 2A and 3A. We find that the characters of all (RH)-vectors l with $0 \leq l(H) \leq q(H) + 1$ come from Riemann surfaces, the smallest representatives w.r.t. the relation $\prec$ are listed in Table 18. The algorithm is not terminated in Step 4. So we compute the set $E = \{2B, 4A\}$, and see that

g	g_0	$l(2A)$	$l(3A)$	$l(2B)$	$l(4A)$		g	g_0	$l(2A)$	$l(3A)$	$l(2B)$	$l(4A)$
13	0	0	0	0	4		10	0	1	0	3	1
10	0	0	0	1	3		7	0	1	0	4	0
7	0	0	0	2	2		9	0	1	1	0	2
4	0	0	0	3	1		6	0	1	1	1	1
13	0	0	0	6	0		3	0	1	1	2	0
3	0	0	1	0	2		19	1	0	0	0	2
18	0	0	1	1	3		16	1	0	0	1	1
12	0	0	1	3	1		13	1	0	0	2	0
9	0	0	1	4	0		9	1	0	1	0	0
8	0	0	2	1	1		24	1	0	1	1	1
5	0	0	2	2	0		25	1	1	0	0	2
19	0	1	0	0	4		22	1	1	0	1	1
16	0	1	0	1	3		19	1	1	0	2	1
13	0	1	0	2	2		15	1	1	1	0	0

TABLE 18. Representatives of $\prec$ for S_4

$b(2B) = q(2B)$ and $b(4A) = q(4A)$. Looking at vectors of norm at most 5 turns out to be sufficient, we find that $V \cup S$ contains no vectors of norm 4 or 5, and we terminate the algorithm with $V = \emptyset$.

The case $g_0 = 1$ is treated analogously. In Step 3, the vector l with $l(H) = 0$ for all $H \in CY(G)/ \sim_G$ lies in S. So we have to proceed with $N = 1$ and $N = 2$, and then terminate with $V = \emptyset$.

Note that the vector $(1, 0, 1, 1, 1)$ does not occur as a representative in Table 18 because the reduction of g_0 is also admissible for the partial ordering $\prec$, so the vector can be reduced to $(0, 0, 1, 1, 1)$.

Since each element in G' is a commutator and G can be generated by two elements, all (RH)-characters with $g_0 \geq 2$ come from Riemann surfaces by Theorem 23.2.

Hence we have shown that an (RH)-character χ of S_4 does not come from a Riemann surface if and only if $\chi = 1_G + n(\chi_4 + \chi_5)$ for a positive integer n.

Note that the vector of norm 3 that corresponds to $(2, 3, 4)$-generation of S_4 does not occur in Table 18, since it does not belong to a character of degree at least 2.   ◇

25. Example: $G = 2^n$

Let $G = \langle a_1, a_2, \ldots, a_n \mid a_i^2, (a_i a_j)^2 \quad \text{for} \quad 1 \le i, j \le n \rangle$, an elementary abelian group of order 2^n. This group is clearly ambivalent. For $a \in G^\times$, we have

$$l_\chi(a) = \frac{r_\chi^*(a)}{2^{n-1}} = \frac{1 - \chi(a)}{2^{n-2}},$$

so the condition (RH) is satisfied if and only if $\chi(a) \le 1$ and $\chi(a) \equiv 1$ (mod 2^{n-2}) for all $a \in G^\times$.

For such a character χ, we define $g_0 = [\chi, 1_G]$, $\Gamma = \Gamma(g_0; 2, 2, \ldots, 2)$, and $\Phi \colon \Gamma \to G$ by mapping $l_\chi(a)$ of the $r = \sum_{a \in G^\times} l_\chi(a)$ elliptic generators to a, for each $a \in G^\times$. We see that χ comes from a Riemann surface if and only if Φ is an epimorphism. By Lemma 23.1, Φ is a homomorphism for any (RH)-character χ.

But we cannot expect that *every* (RH)-character of G comes from a Riemann surface. As in Example 12.5, we may take $1_G + r \cdot \rho_G$, with $1 \le r < n/2 - 1$; we would need a homomorphism of $\Gamma(r + 1; -)$ onto G, but this is impossible since G cannot be generated by $2(r + 1)$ elements.

More general, $\Phi(\Gamma)$ is generated by the set

$$\{\Phi(a_1), \Phi(b_1), \ldots, \Phi(a_{g_0}), \Phi(b_{g_0})\} \cup \{a \in G^\times \mid l_\chi(a) \ne 0\}.$$

So the result is the following.

THEOREM 25.1. *A character χ of the group 2^n comes from a Riemann surface if and only if it satisfies the condition* (RH) *and if $\{a \in G^\times \mid l_\chi(a) \ne 0\}$ generates a subgroup of order at least 2^{n-2g_0}, where $g_0 = [\chi, 1_G]$.*

The surjectivity condition is automatically satisfied in the case $g_0 = 0$ by Lemma 22.19, so a character of the cyclic group of order 2 or of the Klein four group 2^2 comes from a Riemann surface if and only if it satisfies the condition (RH). The same holds for the group 2^3, since the trivial character is the only character for which $g_0 = 1$ and $l_\chi(H) = 0$ for all $H \in CY(G)$, and it does not satisfy the condition (RH).

For $n \ge 4$, each character of the form $1_G + m \cdot (\rho_{\langle a \rangle} - 1_{\langle a \rangle})^G$, with $a \in G^\times$ and m any positive integer, satisfies the condition (RH), but no corresponding homomorphism Φ can be surjective.

The strong symmetric genus of 2^n is zero for $n \le 2$. (Note that the groups $\Gamma(1; -)$ and $\Gamma(0; 2, 2, 2)$ have 2^2 as epimorphic image but do not occur here since they are not Fuchsian. These groups would correspond to the trivial character 1_G and the zero class function, respectively.)

For $n \geq 3$, the strong symmetric genus of 2^n is $1 + 2^{n-2} \cdot (n - 3)$. This is a special case of [**Mac65**, Theorem 4], but we can deduce it also from the observation that a surface kernel homomorphism $\Gamma(g_0; 2, \ldots, 2) \to 2^n$, where Γ has r periods, can be surjective only if either $g_0 = 0$ and $r \geq n+1$ or $g_0 > 0$ and $r \geq n - 2g_0$ holds. In the former case, we get by the Riemann–Hurwitz Formula that the orbit genus of the kernel is $1 - 2^n + 2^{n-2} \cdot r$, with minimum $1 + 2^{n-2} \cdot (n - 3)$, and in the latter case, we get a genus of

$$1 + 2^n \cdot (g_0 - 1) + 2^{n-2} \cdot r \ \geq \ 1 + 2^{n-2} \cdot (4g_0 - 4 + n - 2g_0)$$
$$= \ 1 + 2^{n-2} \cdot (n - 3 + 2g_0 - 1),$$

which is strictly larger. Viewing the group 2^n as an n-dimensional vector space over the field with two elements, we see that any surface kernel epimorphism onto 2^n with kernel of minimal genus maps $r - 1$ generators to linearly independent generators of 2^n, and since the automorphism group of 2^n acts transitively on the bases of 2^n, such an epimorphism is unique up to automorphisms of 2^n. In other words, up to group automorphisms there is a unique character of the group 2^n that comes from a Riemann surface of smallest genus.

26. Example: $G = D_{2n}$

In this section, all characters of dihedral groups $D_{2n} = \langle a, b \mid a^n, b^2, (ab)^2 \rangle$ of order $2n$, with $n \geq 3$, are classified that come from Riemann surfaces. This generalizes the results given in [**Kim93**]. Note that all dihedral groups are ambivalent. We distinguish the cases that n is odd and even.

THEOREM 26.1. *Let χ be an (RH)-character of the dihedral group $G = D_{2n}$ of order $2n$, with $n \geq 3$ and odd. Then χ does not come from a Riemann surface if and only if $[\chi, 1_G] = 1$, $l_\chi(\sigma) = 0$ for all $\sigma \in G \setminus \langle a \rangle$, and $\langle \{a^i \mid l_\chi(a^i) \neq 0\} \rangle \neq \langle a \rangle$.*

Proof. The conjugacy classes of G are $\{1\}$, $\{a^k, a^{-k}\}$ for $1 \leq k \leq (n-1)/2$, and $\{a^k b \mid 0 \leq k \leq n - 1\}$. We have $[b, a^k] = a^{2k}$, so each element in $\langle a \rangle$ is a commutator of elements in G. Because $G/\langle a \rangle$ is abelian, $G' = \langle a \rangle$ and $c.l.(G) = 1$. G can be generated by two elements, so every (RH)-character χ of G with $[\chi, 1_G] \geq 2$ comes from a Riemann surface by Theorem 23.2.

First we compute the matrix T defined in Lemma 22.16. Let 1_G and φ_a denote the linear characters of G, then $\ker(\varphi_a) = \langle a \rangle$. The values T_{H,φ_a} are known by Example 22.20, so let us look at nonlinear $\varphi \in \mathrm{Rat}(G)$. Because $[\varphi, \varphi] = \varphi(1)/2$, we have $T_{H,\varphi} = 4$ if H is the trivial subgroup of G. If H is a nontrivial subgroup of $\langle a \rangle$ then

$$[\psi_{G,H}, \varphi] = [(\rho_H - 1_H)^G, \varphi] = [\rho_H - 1_H, \varphi_H] = \left\{ \begin{array}{ll} 0, & H \subseteq \ker(\varphi), \\ \varphi(1), & H \nsubseteq \ker(\varphi). \end{array} \right.$$

The first case is obvious, the second follows from the fact that the restriction φ_H is the sum of $\varphi(1)$ nontrivial linear characters of H in the second case. This means that

$$T_{H,\varphi} = \begin{cases} 0, & H \subseteq \ker(\varphi), \\ 2, & H \nsubseteq \ker(\varphi), \end{cases}$$

for nontrivial $H = \langle a^k \rangle$. The remaining case $H \nsubseteq \langle a \rangle$ means that H is conjugate to $\langle b \rangle$. Here we get

$$[\psi_{G,\langle b \rangle}, \varphi] = [(\rho_{\langle b \rangle} - 1_{\langle b \rangle})^G, \varphi] = [\rho_{\langle b \rangle} - 1_{\langle b \rangle}, \varphi_{\langle b \rangle}] = \varphi(1)/2,$$

because $\varphi_{\langle b \rangle} = \varphi(1)/2 \cdot \rho_{\langle b \rangle}$, and $\rho_{\langle b \rangle} - 1_{\langle b \rangle}$ is the unique nontrivial irreducible character of $\langle b \rangle$. So we get $T_{\langle b \rangle, \varphi} = 1$.

Putting the pieces together, the matrix T looks as follows:

$$\begin{pmatrix} 2 & 2 & 4 & \cdots & 4 \\ \hline 0 & 0 & & & \\ \vdots & \vdots & & 0 \text{ or } 2 & \\ 0 & 0 & & & \\ \hline 0 & 1 & 1 & \cdots & 1 \end{pmatrix}.$$

The first two columns belong to the characters 1_G and φ_a, and the others to nonlinear characters. The first row belongs to the trivial subgroup, the second portion of rows to nontrivial subgroups of $\langle a \rangle$, and the last row belongs to $\langle b \rangle$.

Now let χ be an (RH)-character of G, and set $g_0 = [\chi, 1_G]$. We want to construct $\Phi \in \text{Epi}_{L(\chi)}(\Gamma(\chi), G)$ in all cases for which Theorem 26.1 claims that such epimorphisms exist.

From the above matrix, we read off that

$$2[\chi, \varphi_a] = 2(g_0 - 1) + l_\chi(b),$$

so $l_\chi(b)$ is even.

If $l_\chi(b) \geq 4$ then we choose arbitrary images of the hyperbolic generators a_i, b_i of $\Gamma(\chi)$ and of those elliptic generators c_j that are mapped to elements of $\langle a \rangle$. Let

$$a^k = \prod_{i=1}^{g_0} [\Phi(a_i), \Phi(b_i)] \cdot \prod_{\{j | \Phi(c_j) \in \langle a \rangle\}} \Phi(c_j).$$

Now choose two images $a^{-k}b$, b, and $l_\chi(b) - 2$ images ab for the remaining elliptic generators. This defines an epimorphism as desired.

If $l_\chi(b) = 2$ and $g_0 > 0$ then we choose a and b as images of the hyperbolic generators, arbitrary images in $\langle a \rangle$, and $a^{-k}b$, b as above, for the appropriate value k.

In the remaining case that $l_\chi(b) = 0$ or $g_0 = 0$, we define

$$U = \langle\{a^i \mid 1 \le i \le n-1, l_\chi(a^i) \ne 0\}\rangle.$$

If $g_0 = 0$ then $l_\chi(b) \ne 0$ because $0 \le 2 \cdot [\chi, \varphi_a] = -2 + l_\chi(b)$, and we get a homomorphism Φ as in the situations above. It is surjective if $U = \langle a \rangle$, and we show that this is automatically satisfied. The kernels of the nonlinear characters in $\mathrm{Rat}(G)$ are exactly the proper subgroups of $\langle a \rangle$, so if U is a proper subgroup of $\langle a \rangle$ we can take $\varphi \in \mathrm{Rat}(G)$ with $U \le \ker(\varphi)$, and we get

$$2 \cdot [\chi, \varphi] = -4 + \sum_{\substack{H \in CY(\langle a \rangle) \\ H \subseteq \ker(\varphi)}} l_\chi(H) \cdot T_{H,\varphi} + \sum_{\substack{H \in CY(\langle a \rangle) \\ H \not\subseteq \ker(\varphi)}} l_\chi(H) \cdot T_{H,\varphi} + l_\chi(b) \cdot T_{\langle b \rangle, \varphi}.$$

For each summand of the first sum, $T_{H,\varphi} = 0$, and $l_\chi(H) = 0$ for each summand in the second sum. Since $T_{\langle b \rangle, \varphi} = 1$, we see that $l_\chi(b) \ge 4$, contrary to our assumptions.

The last case we have to deal with is that $l_\chi(b) = 0$ and $g_0 = 1$. If $U = \langle a \rangle$ then we choose arbitrary images for the elliptic generators, and denote their product by a^k. Let $a^k = a^{-2j}$ for an integer j, and choose $a^j b$ and b as images of the hyperbolic generators. Because of $[a^j b, b] = a^{2j}$, this yields that Φ is an epimorphism.

It remains to show that there is *no* such epimorphism in the case that $g_0 = 1$, $U \ne \langle a \rangle$, and $l_\chi(b) = 0$. But this follows from Theorem 23.8.   □

THEOREM 26.2. *Let χ be an (RH)-character of the dihedral group $G = D_{4n}$ of order $4n$, with $n \ge 2$. Then χ does not come from a Riemann surface if and only if $[\chi, 1_G] = 1$, $l_\chi(\sigma) = 0$ for all $\sigma \in G \setminus \langle a \rangle$, and $U = \langle\{a^i \mid l_\chi(a^i) \ne 0\}\rangle$ satisfies one of the following conditions.*

(a) *U is different from $\langle a \rangle$ and $\langle a^2 \rangle$, or*
(b) *$U = \langle a^2 \rangle$, the 2-part n_2 of n is larger than 1, and*

$$\sum_{|a^{2i}| \equiv 0 \pmod{n_2}} l_\chi(a^{2i})$$

is even.

Proof. The conjugacy classes of G are $\{1\}$, $\{a^n\}$, $\{a^k, a^{-k}\}$ for $1 \le k \le n-1$, $\{a^{2k}b \mid 0 \le k \le n-1\}$, and $\{a^{2k+1}b \mid 0 \le k \le n-1\}$. We have $G' = \langle a^2 \rangle$, $c.l.(G) = 1$, and every (RH)-character χ of G with $[\chi, 1_G] \ge 2$ comes from a Riemann surface by Theorem 23.2.

The matrix T of Lemma 22.16 is computed as in the proof of Theorem 26.1. Let 1_G, φ_a, φ_b, and φ_{ab} denote the linear characters of G, with $\ker(\varphi_\sigma) = \langle a^2, \sigma \rangle$. The values T_{H,φ_σ} are known by Example 22.20, and for nonlinear

$\varphi \in \mathrm{Rat}(G)$, we have $[\varphi, \varphi] = \varphi(1)/2$ and thus $T_{H,\varphi} = 4$ if H is the trivial subgroup of G, and

$$T_{H,\varphi} = \begin{cases} 0, & H \subseteq \ker(\varphi), \\ 2, & H \not\subseteq \ker(\varphi), \end{cases}$$

if H is a nontrivial subgroup of $\langle a \rangle$. For $H \not\subseteq \langle a \rangle$, H is conjugate to one of $\langle b \rangle$, $\langle ab \rangle$. Here we get

$$[\psi_{G,H}, \varphi] = [(\rho_H - 1_H)^G, \varphi] = [\rho_H - 1_H, \varphi_H] = \varphi(1)/2,$$

because $\varphi_H = \varphi(1)/2 \cdot \rho_H$, and $\rho_H - 1_H$ is the unique nontrivial irreducible character of H. This implies $T_{H,\varphi} = 1$.

The matrix T looks as follows:

$$\begin{pmatrix} 2 & 2 & 2 & 2 & 4 & \cdots & 4 \\ \hline 0 & 0 & 0 & 0 & & & \\ \vdots & \vdots & \vdots & \vdots & 0 & \text{or} & 2 \\ 0 & 0 & 0 & 0 & & & \\ \hline 0 & 0 & 1 & 1 & & & \\ \vdots & \vdots & \vdots & \vdots & 0 & \text{or} & 2 \\ 0 & 0 & 1 & 1 & & & \\ \hline 0 & 1 & 0 & 1 & 1 & \cdots & 1 \\ 0 & 1 & 1 & 0 & 1 & \cdots & 1 \end{pmatrix}$$

The first two columns belong to the characters 1_G and φ_a, the next two columns to φ_b and φ_{ab}, and the others to nonlinear characters. The first row belongs to the trivial subgroup, the second portion of rows to nontrivial subgroups of $\langle a^2 \rangle$, the third portion to the remaining subgroups of $\langle a \rangle$, and the last two rows belong to $\langle b \rangle$ and $\langle ab \rangle$.

Now let χ be an (RH)-character of G, and set $g_0 = [\chi, 1_G]$. We want to construct $\Phi \in \mathrm{Epi}_{L(\chi)}(\Gamma(\chi), G)$ in all cases for which Theorem 26.2 claims that such epimorphisms exist.

The column of φ_a in the above matrix yields that $l_\chi(b) \equiv l_\chi(ab) \pmod 2$.

If both $l_\chi(b)$ and $l_\chi(ab)$ are odd then we choose arbitrary images for the hyperbolic generators a_i, b_i of $\Gamma(\chi)$ and for those of the elliptic generators c_j that must be mapped to powers of a. Let

$$a^k = \prod_{i=1}^{g_0} [\Phi(a_i), \Phi(b_i)] \cdot \prod_{\Phi(c_j) \in \langle a \rangle} \Phi(c_j).$$

Because

$$2[\chi, \varphi_b] = 2(g_0 - 1) + l_\chi(b) + \sum_{\substack{H \subseteq \langle a \rangle \\ H \not\subseteq \langle a^2 \rangle}} l_\chi(H),$$

the exponent k is odd, hence we may choose one image $a^k b$, $l_\chi(ab) - 1$ images ab, and $l_\chi(b)$ images b. This defines a homomorphism $\Phi \in \mathrm{Hom}_{L(\chi)}(\Gamma(\chi), G)$. For $g_0 > 0$, we can choose the images of the hyperbolic generators such that Φ is surjective. If $g_0 = 0$ then we have no such choice, but we claim that Φ is automatically surjective. To prove this, it is sufficient to show that the images inside $\langle a \rangle$ generate $\langle a \rangle$. So we set $U = \langle \{ a^j \mid 1 \leq j < 2n, l_\chi(a^j) \neq 0 \} \rangle$, and suppose that $U \neq \langle a \rangle$. Clearly $U \not\subseteq \langle a^2 \rangle$ by consideration of φ_b, so there is a nonlinear character $\varphi \in \mathrm{Rat}(G)$ with $\ker(\varphi) = U$, namely the Galois sum of the irreducible character ϑ^G of G, where $\vartheta \in \mathrm{Irr}(\langle a \rangle)$ has kernel $\langle a^m \rangle$. The multiplicity of φ as a constituent of $\chi + \overline{\chi}$ is

$$-4 + l_\chi(b) \cdot T_{\langle b \rangle, \varphi} + l_\chi(ab) \cdot T_{\langle ab \rangle, \varphi} = -4 + l_\chi(b) + l_\chi(ab),$$

because for each nontrivial subgroup H of $\langle a \rangle$, either $l_\chi(H)$ or $T_{H,\varphi}$ is zero. We see that $l_\chi(b) + l_\chi(ab) \geq 4$, hence one of $l_\chi(b)$, $l_\chi(ab)$ is at least 3. If $l_\chi(ab) \geq 3$ then there are generator images b and ab, thus a is in the image of Φ. If $l_\chi(b) \geq 3$ then we can choose two images $a^{k-1}b$ instead of b, and again have a in the image of Φ.

Now consider the case that both $l_\chi(b)$ and $l_\chi(ab)$ are even. As above, we choose images for the hyperbolic generators and for those of the elliptic generators that must be mapped to powers of a, and define a^k. Here the column of φ_b in T guarantees that k is even. In the following, we distinguish several cases, according to whether $l_\chi(b)$ or $l_\chi(ab)$ is zero.

If $l_\chi(b) \neq 0$ then we choose one image $a^{-k}b$, $l_\chi(b) - 1$ images b, and $l_\chi(ab)$ images ab. This defines an element in $\mathrm{Hom}_{L(\chi)}(\Gamma(\chi), G)$, which is automatically surjective if $l_\chi(ab) \neq 0$, and which can be chosen as an epimorphism in the case $g_0 \neq 0$.

So for $l_\chi(b) \neq 0$, the case that $g_0 = l_\chi(ab) = 0$ remains. The column of φ_b guarantees that

$$U = \langle \{ a^j \mid 1 \leq j < 2n, l_\chi(a^j) \neq 0 \} \rangle$$

is not contained in $\langle a^2 \rangle$, and as above, there is a nonlinear $\varphi \in \mathrm{Rat}(G)$ with $\ker(\varphi) = \langle a^m \rangle$ for which $a_\varphi = -4 + l_\chi(b)$. So $l_\chi(b) \geq 4$, and we can choose two images $a^2 b$ instead of b. Then a^2 is in the image of Φ, and thus Φ is surjective.

If $l_\chi(ab) \neq 0$ then we proceed analogously, that is, we choose appropriate images of the hyperbolic generators if $g_0 \neq 0$, and if $g_0 = l_\chi(b) = 0$, we get $l_\chi(ab) \geq 4$.

In the remaining case that $l_\chi(b) = l_\chi(ab) = 0$, we have necessarily $g_0 \neq 0$ by the column of φ_a. So let $g_0 = 1$. Here we choose arbitrary images for the elliptic generators c_i, and define a^{2k} to be the product of these images; note that the exponent is even by Lemma 23.1.

Let U be defined as above. If $U = \langle a \rangle$ then we may choose b and a^{-k} as images of the hyperbolic generators, and get an epimorphism. If U is a proper subgroup of $\langle a^2 \rangle$ then no such epimorphism exists, by Theorem 23.8.

If U is a proper subgroup of $\langle a \rangle$ but not contained in $\langle a^2 \rangle$ then we apply a slight modification of the argument of Theorem 23.8. That is, if x and y are elements in G with $[x, y] \in U$ then the derived subgroup of $\langle U, x, y \rangle$ is the normal closure of the group generated by $[U, U]$, $[U, x]$, $[U, y]$, and $[x, y]$; but $[U, U]$ is trivial, every commutator $[u, x]$ with $u \in U$ lies in $U \cdot U = U \cap \langle a^2 \rangle$, and also $[U, y]$ and $[x, y]$ lie in $U \cap \langle a^2 \rangle$. Hence $\langle U, x, y \rangle' \subseteq U \cap \langle a^2 \rangle$, which means that $\langle U, x, y \rangle \neq G$.

Only the case $U = \langle a^2 \rangle$ is left, and we have to show that an epimorphism exists if and only if n is odd or $\sum_{|a^{2i}| \equiv 0 \;(\mathrm{mod}\; n_2)} l_\chi(a^{2i})$ is odd. These conditions are equivalent to the condition that the index $[\langle a^2 \rangle : \langle a^k \rangle]$ is odd, where a^k is the product of images of the elliptic generators; note that this is independent of the choice of the images.

If the index is odd then there is an odd integer j such that $a^k = a^{2j}$, and we may choose a^j and b as images of the hyperbolic generators. If the index is even then we apply the above modification of Theorem 23.8 again; $[x, y] = a^k$ lies in $\langle a^4 \rangle$ for any x and y in G, the same holds for $[U, x]$, $[U, y]$, and $[U, U]$, so $\langle U, x, y \rangle' \subseteq \langle a^4 \rangle$ proves that $\langle U, x, y \rangle \neq G$.   □

REMARK 26.3. The exceptions in Theorem 26.2 are covered by the results of Section 23.2. That is, any U different from $\langle a \rangle$ and $\langle a^2 \rangle$ is either a proper subgroup of $\langle a^2 \rangle = G'$ or contains a normal subgroup N of index 2 such that $F = G/N$ is a nonabelian (in fact dihedral) group. So we can apply Theorem 23.5 in the first case, and Theorem 23.7 in the second. In situation (b), again Theorem 23.7 can be applied, with $N = \langle a^4 \rangle$.

Both the statement and the proof of Theorem 26.2 are easier for the special case that the order of G is a power of 2.

COROLLARY 26.4. *An* (RH)-*character* χ *of* $G = D_{2^n}$, *with* $n \geq 3$, *does not come from a Riemann surface if and only if* $[\chi, 1_G] = 1$, $l_\chi(b) = l_\chi(ab) = l_\chi(a) = 0$, *and* $l_\chi(a^2) \equiv 0 \pmod 2$.

Proof. The statement is obtained from that of Theorem 26.2 using the fact that the subgroups of the cyclic group $\langle a \rangle$ of order 2^{n-1} form a chain. So the condition (a) in Theorem 26.2 means $l_\chi(a) = l_\chi(a^2) = 0$, and the condition (b) means that $l_\chi(a) = 0$ and $l_\chi(a^2)$ is nonzero and even.   □

REMARK 26.5. Note that the matrix T has a particularly easy structure for dihedral groups of 2-power order. That is, the third portion of rows consists of the single row for the group $\langle a \rangle$, which has values 2 in all columns corresponding to nonlinear characters. And the submatrix formed by the second portion of rows and the third portion of columns is a triangular matrix if the rows and columns are sorted according to containment of kernels and subgroups, respectively.

Also it is easy to show that every (RH)-character χ of D_{2^n} with $[\chi, 1_G] = 0$ comes from a Riemann surface. The argument is that in any 2-group, all maximal subgroups are normal, and hence each proper subgroup is contained in a proper *normal* subgroup. So we can apply Lemma 22.19.

EXAMPLE 26.6 (see [**Kim93**, Theorem 3.1]). The matrices T for the groups D_8 and Q_8 are shown in Table 19; the rows and columns are ordered in the same way as the columns and rows in Table 15 on page 108. The statement of Corollary 26.4 for the case that $G = D_8$ means that the only G-characters that satisfy the condition (RH) but do not come from Riemann surfaces are the characters $1_G + n \cdot (\rho_{\langle z \rangle} - 1_{\langle z \rangle})^G = 1_G + 2n \cdot \chi_5$, which we have met already in Example 23.6.

$$
D_8 : \begin{pmatrix} 2 & 2 & 2 & 2 & 4 \\ 0 & 0 & 0 & 0 & 2 \\ 0 & 0 & 1 & 1 & 2 \\ 0 & 1 & 0 & 1 & 1 \\ 0 & 1 & 1 & 0 & 1 \end{pmatrix}
\qquad
Q_8 : \begin{pmatrix} 2 & 2 & 2 & 2 & 4 \\ 0 & 0 & 0 & 0 & 2 \\ 0 & 0 & 1 & 1 & 2 \\ 0 & 1 & 0 & 1 & 2 \\ 0 & 1 & 1 & 0 & 2 \end{pmatrix}
$$

TABLE 19. Matrices T for D_8 and Q_8

The same holds for the quaternion group Q_8. Note that the arguments used in the proof of Theorem 26.2 hold also for Q_8, most of them simply because the columns in the matrices T that correspond to linear characters of D_8 and Q_8 are equal.   ◇

27. Example: $G = L_2(2^n)$

In this section, let G be the projective special linear group $L_2(q)$, for $q = 2^n$. The case $L_2(2) \cong D_6$ has been treated already in Section 21, so we assume $n \geq 2$ in the following.

27.1. Finiteness Result.
The irreducible characters of G can be found in [**Sch07**, p. 134], they are shown in Table 20; the parameter values are $1 \leq a, \alpha \leq \frac{q-2}{2}$ and $1 \leq b, \beta \leq \frac{q}{2}$. Note that G is ambivalent.

THEOREM 27.1. *Let $G = L_2(2^n)$, with $n \geq 2$. Then there are only finitely many G-characters that satisfy the condition* (RH) *but do not come from Riemann surfaces.*

	1	P	A^a	B^b
1_G	1	1	1	1
φ_β	$q-1$	-1	0	$-\zeta_{q+1}^{-b\beta} - \zeta_{q+1}^{-b\beta}$
ψ	q	0	1	-1
μ_α	$q+1$	1	$\zeta_{q-1}^{a\alpha} + \zeta_{q-1}^{-a\alpha}$	0

TABLE 20. Irreducible Characters of $L_2(2^n)$

Proof. We want to apply Theorem 23.13. So we must show that $rep(\langle P \rangle^G)$ $< \infty$; note that the elements in classes different from that of P have odd order. In order to show that $\mathrm{Hom}_{(P^G, P^G, (P^{-1})^G)}(0, G) \neq \emptyset$, by Theorem 14.1 it suffices to prove that

$$S(\sigma) = \sum_{\chi \in \mathrm{Irr}(G)} \frac{\chi(\sigma)^2 \overline{\chi(\sigma)}}{\chi(1)}$$

is nonzero for $\sigma = P$. Note that we can omit the complex conjugation because G is ambivalent.

We compute

$$S(P) = 1 + \frac{q}{2} \cdot \frac{-1}{q-1} + \frac{q-2}{2} \cdot \frac{1}{q+1} = \left(1 - \frac{q}{2(q-1)}\right) + \frac{q-2}{2(q+1)},$$

which is nonzero because both summands are positive. $\qquad\square$

One can show that $\mathrm{Hom}_{(\sigma^G, \sigma^G, \sigma^G)}(0, G) \neq \emptyset$ for all $\sigma \in G$, that is, each $\sigma \in G$ is the product of two elements that are conjugate in G to σ. Before we compute $S(A^a)$ and $S(B^b)$, we need a preparatory lemma.

LEMMA 27.2. *Let a and m be positive integers, and $a < m$.*

(i)

$$\sum_{i=1}^{m-1} \zeta_m^{ai} = -1 + \sum_{i=1}^{m} \zeta_{m/\gcd(m,a)}^{ai/\gcd(m,a)} = \begin{cases} m-1, & \text{if } m \text{ divides } a, \\ -1, & \text{otherwise.} \end{cases}$$

(ii) *If m is odd then*

$$\sum_{i=1}^{(m-1)/2} (\zeta_m^{ai} + \zeta_m^{-ai})^3 = \begin{cases} m-4, & \text{if } m \text{ divides } 3a, \\ -4, & \text{otherwise.} \end{cases}$$

(iii) *If m is even then*

$$\sum_{i=1}^{m/2-1} (\zeta_m^{ai} + \zeta_m^{-ai})^3 = \begin{cases} m - 4(1 + (-1)^a), & \text{if } m \text{ divides } 3a, \\ -4(1 + (-1)^a), & \text{otherwise.} \end{cases}$$

Proof. In (i), the first equality is obvious, and the second holds because in the summation from 1 to m, each $(m/\gcd(m,a))$-th root of unity occurs with multiplicity $\gcd(m,a)$, and the sum over all k-th roots of unity is zero if $k \neq 1$.

For (ii), note that

$$\sum_{i=1}^{(m-1)/2} (\zeta_m^{ai} + \zeta_m^{-ai})^3 = \sum_{i=1}^{m-1} \zeta_m^{3ai} + 3\sum_{i=1}^{m-1} \zeta_m^{ai},$$

where the first sum evaluates to $m-1$ or -1, by part (i), and analogously the second sum evaluates to -1 because m does not divide a.

In the same way, (iii) follows from

$$\sum_{i=1}^{m/2-1} (\zeta_m^{ai} + \zeta_m^{-ai})^3 = \sum_{i=1}^{m-1} \zeta_m^{3ai} - \zeta_m^{3am/2} + 3\sum_{i=1}^{m-1} \zeta_m^{ai} - 3\zeta_m^{am/2},$$

the values of the sums are $m-1$ or -1 by (i), and $\zeta_m^{3am/2} = \zeta_m^{am/2} = (-1)^a$. $\square$

Now we get

$$S(A^a) = 1 + \frac{1}{q} + \frac{1}{q+1} \cdot \sum_{\alpha=1}^{(q-2)/2} (\zeta_{q-1}^{a\alpha} + \zeta_{q-1}^{-a\alpha})^3$$

$$\geq 1 + \frac{1}{q} - \frac{4}{q+1} > 1 - \frac{4}{q+1} > 0$$

because $q \geq 4$. Analogously,

$$S(B^b) = 1 - \frac{1}{q-1} \cdot \sum_{\beta=1}^{q/2} (\zeta_{q+1}^{b\beta} + \zeta_{q+1}^{-b\beta})^3 - \frac{1}{q}$$

$$\geq 1 - \frac{q-3}{q-1} - \frac{1}{q} > 1 - \frac{q-3}{q-1} - \frac{1}{q-1} = 1 - \frac{q-2}{q-1} > 0.$$

REMARK 27.3. The above calculations imply the known fact that $c.l.(G) = 1$. That is, if $\sigma = \sigma^\tau \sigma^\kappa$ is a product of two of its conjugates then $\sigma^\kappa = (\sigma^\tau)^{-1}\sigma = [\tau, \sigma]$ is a commutator.

27.2. An Exceptional Character. For each value of $n \geq 2$, we construct an (RH)-character of $L_2(2^n)$ that does not come from a Riemann surface. We need the following technical lemma.

LEMMA 27.4. *Let $G = L_2(q)$, with $q = 2^n$ and $n \geq 2$, and $a, b \in G$ be such that the commutator $[a, b]$ is an involution. Then $\langle a, b \rangle$ is a subgroup of $N_G(S)$, for a Sylow 2-subgroup S of G.*

Proof. By Theorem 14.1 and the character table of G shown in Table 20, we have $|G| = q \cdot (q^2 - 1)$ and $|P^G| = q^2 - 1$, thus

$$
\begin{aligned}
|\mathrm{Hom}_{(P^G)}(1, G)| &= |G| \cdot |P^G| \cdot \sum_{\chi \in \mathrm{Irr}(G)} \frac{\chi(P)}{\chi(1)} \\
&= q \cdot (q^2 - 1)^2 \cdot \left(1 + \frac{q}{2} \cdot \frac{-1}{q-1} + \frac{q-2}{2} \cdot \frac{1}{q+1}\right) \\
&= q^2 \cdot (q^2 - 1) \cdot (q - 2).
\end{aligned}
$$

Now we show that the image of each element in $\mathrm{Hom}_{(P^G)}(1, G)$ is contained in a subgroup $N = N_G(S)$, for $S \in Syl_2(G)$. Since all elements of even order in G are involutions, S is elementary abelian. N/S is a cyclic group of order $q - 1$, and the centralizer $C_N(a) = C_G(a)$ is a complement of S in N for each element $a \in N \setminus S$, see for example [**Hup83**, Sätze II.8.2 and II.8.3]. So the commutator $[a, b]$ of two elements $a, b \in N$ lies in S, a and b commute if and only if either both lie in S or both lie in the same cyclic subgroup of order $q - 1$ in N, and if a and b do not commute then $[a, b] \in S^\times \subseteq P^G$. The intersection of two different Sylow 2-subgroups of G is trivial because each element in $S^\times$ is central in S and has centralizer order $q = |S|$.

This means that for each $a \in P^G$, there is a unique $S \in Syl_2(G)$ that contains a, and for each $b \in N_G(S) \setminus S$, the commutator $[a, b]$ is an involution; hence

$$
|P^G| \cdot |N_G(S) \setminus S| = (q^2 - 1)(q(q-1) - q) = q(q^2 - 1)(q - 2)
$$

elements in $\mathrm{Hom}_{(P^G)}(1, G)$ arise this way.

Analogously, we get the same number of homomorphisms by taking $b \in P^G$ and $a \in N_G(S) \setminus S$.

The third possibility for constructing elements in $\mathrm{Hom}_{(P^G)}(1, G)$ is to take $S \in Syl_2(G)$, and two noncommuting elements $a, b \in N_G(S) \setminus S$. For each choice of S, there are $|N \setminus S| = q(q-2)$ possibilities for a, and $q(q-1) - q - (q-2) = (q-2)(q-1)$ possibilities for b. Because $[a, b] \in S^\times$, a and b determine a unique $S \in Syl_2(G)$, so we get $(q+1) \cdot q(q-2) \cdot (q-2)(q-1) = q(q^2 - 1)(q-2)^2$ new elements in $\mathrm{Hom}_{(P^G)}(1, G)$.

Taking the union with the ones for which a or b is an involution, we have found

$$
2q(q^2 - 1)(q - 2) + q(q^2 - 1)(q - 2)^2 = q^2(q^2 - 1)(q - 2)
$$

homomorphisms. $\qquad\qquad\qquad\qquad\qquad\qquad\qquad\qquad\qquad\qquad\quad$ $\square$

Now we compute the values $T_{H,\varphi}$ for $H = \langle P \rangle$. We have $\psi_{G,H} = \vartheta^G$, where ϑ is the unique nontrivial irreducible character of H. Trivially $T_{H,1_G} = 0$, and for $\varphi \in \mathrm{Rat}(G) \setminus \{1_G\}$,

$$[\psi_{G,H}, \varphi] = [\vartheta, \varphi_H] = \frac{1}{2}(\varphi(1) - \varphi(P)) = \frac{q}{2}[\varphi, \varphi],$$

and thus $T_{H,\varphi} = q/2$ holds.

So the G-character

$$\chi = 1_G + \frac{q}{4} \sum_{\varphi \in \mathrm{Irr}(G) \setminus \{1_G\}} \varphi = 1_G + \frac{q}{4} \sum_{\varphi \in \mathrm{Rat}(G) \setminus \{1_G\}} \varphi$$

satisfies the condition (RH), with $g_0 = 1$, $l_\chi(P) = 1$, and $l_\chi(H) = 0$ for $H \in CY(G)$ with $H \not\leq_G \langle P \rangle$. But χ does not come from a Riemann surface because $\mathrm{Epi}(\Gamma'(\chi), G) = \mathrm{Epi}_{\langle PG \rangle}(\Gamma(1; 2), G) = \emptyset$ by Lemma 27.4.

27.3. Small Values of n. We look at $G = L_2(2^n)$ for small values of $2 \leq n \leq 5$. Because $c.l.(G) = 1$ (see Remark 27.3) and because G can be generated by two elements (see for example [**Hup83**, II. 8.]), Corollary 23.3 yields that only $g_0 = [\chi, 1_G] \in \{0, 1\}$ must be considered.

The group $G = L_2(4)$ is isomorphic with the alternating group A_5. Table 21 lists, for $g_0 = 0$ and $g_0 = 1$, in the first three columns the entries of some vectors $(l_\chi(2A), l_\chi(3A), l_\chi(5AB))$. The sign in the fourth column is "+" if the vector describes an (RH)-character χ, and "−" otherwise. The degree $g = \chi(1)$ is given in column 5, and the remaining column(s) describe(s) $\Phi \in \mathrm{Epi}_{L(\chi)}(\Gamma(\chi), G)$ if such an epimorphism exists. For any (RH)-character χ of G, the corresponding vector l_χ either is contained in the table or can be obtained from a vector in the table for which Φ exists by increasing some of the entries. Thus we see that only one (RH)-character of G does not come from a Riemann surface. Of course this is the character of degree 16 that has been considered already in Example 22.12 and in Section 27.2.

Analogous computations yield that $L_2(8)$ and $L_2(32)$ also have exactly one (RH)-character that does not come from a Riemann surface, the character known from Section 27.2.

For $L_2(16)$, the situation is different. This group has exactly the following four (RH)-characters that do not come from Riemann surfaces. (We write $\chi_{n,\ldots,m}$ for $\chi_n + \chi_{n+1} + \cdots + \chi_m$.)

g	g_0	Nonzero $l_\chi(\sigma)$	χ
273	0	$l_\chi(3A) = 2, l_\chi(5A) = 1$	$\chi_{2,\ldots,9} + \chi_{11} + 2\chi_{14,\ldots,17}$
545	0	$l_\chi(3A) = 1, l_\chi(5A) = 2$	$2\chi_{2,\ldots,9} + \chi_{10} + \chi_{11} + 2\chi_{12,13} + 3\chi_{14,\ldots,17}$
817	0	$l_\chi(5A) = 3$	$3\chi_{2,\ldots,9} + 2\chi_{10} + \chi_{11} + 4\chi_{12,13} + 4\chi_{14,\ldots,17}$
1 021	1	$l_\chi(2A) = 1$	$1_G + 4\chi_{2,\ldots,17}$

Note that the three characters χ with $[\chi, 1_G] = 0$ are our first examples of (RH)-characters with this property that do not come from Riemann surfaces.

$g_0 = 0$:

$l_\chi(2A)$	$l_\chi(3A)$	$l_\chi(5AB)$	(RH) ?	g	$\Phi(c_1),\ldots,\Phi(c_r)$
0	0	0	−		
0	0	1	−		
0	0	2	−		
0	0	3	+	13	(1,2,3,4,5), (1,4,5,2,3), (1,2,4,5,3)
0	1	0	−		
0	1	1	−		
0	1	2	+	9	(3,4,5), (1,5,3,2,4), (1,3,4,2,5)
0	2	0	−		
0	2	1	+	5	(3,4,5), (1,2,4), (1,3,5,4,2)
0	3	0	−		
0	4	0	+	21	(3,4,5), (1,3,2), (1,2,3), (3,5,4)
1	0	0	−		
1	0	1	−		
1	0	2	+	4	(2,3)(4,5), (1,5,2,4,3), (1,2,4,3,5)
1	1	0	−		
1	1	1	−	0	(2,3)(4,5), (1,5,3), (1,2,3,4,5)
1	2	0	−		
1	3	0	+	16	(2,3)(4,5), (1,5,3), (1,2,5), (3,4,5)
2	0	0	−		
2	0	1	−		
2	1	0	−		
2	2	0	+	11	(2,3)(4,5), (1,3)(2,5), (1,2,3), (3,4,5)
3	0	0	−		
3	0	1	+	10	(2,3)(4,5), (1,3)(2,4), (2,4)(3,5), (1,2,3,4,5)
3	1	0	+	6	(2,3)(4,5), (1,4)(2,3), (1,4)(3,5), (3,4,5)

$g_0 = 1$:

$l_\chi(2A)$	$l_\chi(3A)$	$l_\chi(5AB)$	(RH) ?	g	$\Phi(a_1),\Phi(b_1)$	$\Phi(c_1),\ldots,\Phi(c_r)$
0	0	0	+	1		
0	0	1	+	25	(2,3)(4,5), (1,2,4)	(1,4,2,3,5)
0	1	0	+	21	(2,3)(4,5), (1,2,5,3,4)	(1,2,3)
0	1	1	+	45	(2,3)(4,5), (1,2,4)	(1,5,4), (1,2,3,5,4)
1	0	0	+	16		
2	0	0	+	31	(2,3)(4,5), (1,2,5,3,4)	(1,3)(4,5), (2,3)(4,5)
1	0	1	+	40	(2,3)(4,5), (1,2,5,3,4)	(2,4)(3,5), (1,2,4,3,5)
1	1	0	+	36	(2,3)(4,5), (1,2,5,3,4)	(1,3)(2,4), (2,4,3)
1	1	1	+	60	(2,3)(4,5), (1,2,5,3,4)	(1,3)(2,4), (1,5,2,4,3), (1,2,5)

TABLE 21. Surface Kernel Epimorphisms for A_5

28. Example: $G = L_2(q)$ for $q \equiv 1$ (mod 4)

In this section, let G be the projective special linear group $L_2(q)$, for $q \equiv 1$ (mod 4), $q \geq 5$.

28.1. Finiteness Result. The irreducible characters of G can be found in [**Sch07**, p. 128], they are shown in Table 22; the parameter values are $1 \leq \alpha \leq \frac{q-5}{4}$ and $1 \leq a, b, \beta \leq \frac{q-1}{4}$. Note that G is ambivalent.

	1	P	Q	A^a	B^b
1_G	1	1	1	1	1
φ_β	$q-1$	-1	-1	0	$-\zeta_{(q+1)/2}^{b\beta} - \zeta_{(q+1)/2}^{-b\beta}$
ψ	q	0	0	1	-1
μ_α	$q+1$	1	1	$\zeta_{(q-1)/2}^{a\alpha} + \zeta_{(q-1)/2}^{-a\alpha}$	0
θ_1	$\frac{q+1}{2}$	$\frac{1+\sqrt{q}}{2}$	$\frac{1-\sqrt{q}}{2}$	$(-1)^a$	0
θ_2	$\frac{q+1}{2}$	$\frac{1-\sqrt{q}}{2}$	$\frac{1+\sqrt{q}}{2}$	$(-1)^a$	0

TABLE 22. Irreducible Characters of $L_2(q)$ for $q \equiv 1 \pmod 4$

THEOREM 28.1. *Let $G = L_2(q)$, with $q \equiv 1 \pmod 4$, $q \geq 5$. Then there are only finitely many G-characters that satisfy the condition* (RH) *but do not come from Riemann surfaces.*

Proof. The elements of even order in G lie in $\langle A^a \rangle$, $\langle B^b \rangle$. As in the proof of Theorem 27.1, it suffices to show that $S(A^a)$ and $S(B^b)$ are nonzero. Using Lemma 27.2, we compute

$$
\begin{aligned}
S(A^a) &= 1 + \frac{1}{q} + \frac{1}{q+1} \sum_{\alpha=1}^{(q-5)/4} (\zeta_{(q-1)/2}^{a\alpha} + \zeta_{(q-1)/2}^{-a\alpha})^3 + 2 \cdot \frac{(-1)^a}{(q+1)/2} \\
&\geq 1 + \frac{1}{q} + \frac{-4 \cdot (1 + (-1)^a)}{q+1} + \frac{4 \cdot (-1)^a}{q+1} = 1 + \frac{1}{q} - \frac{4}{q+1} > 0
\end{aligned}
$$

and

$$
\begin{aligned}
S(B^b) &= 1 - \frac{1}{q-1} \sum_{\beta=1}^{(q-1)/4} (\zeta_{(q+1)/2}^{b\beta} + \zeta_{(q+1)/2}^{-b\beta})^3 - \frac{1}{q} \\
&\geq 1 - \frac{(q+1)/2 - 4}{q-1} - \frac{1}{q} \\
&= \left(1 - \frac{q+1}{2(q-1)}\right) + \left(\frac{4}{q-1} - \frac{1}{q}\right) > 0.
\end{aligned}
$$

$\square$

REMARK 28.2. One can show that $\mathrm{Hom}_{(\sigma G, \sigma G, \sigma G)}(0, G) \neq \emptyset$ for all $\sigma \in G$. That is, $S(P) =$

$$1 + \frac{q-1}{2} \cdot \frac{-1}{q-1} + \frac{q-5}{4} \cdot \frac{1}{q+1} + \frac{2}{q+1}\left(\left(\frac{1+\sqrt{q}}{2}\right)^3 + \left(\frac{1-\sqrt{q}}{2}\right)^3\right)$$

$$= \frac{3}{4} + \frac{1}{4(q+1)}(q - 5 + 2(1+3q)) > 0,$$

and $S(Q) = S(P)$. As noted in Remark 27.3, we have thus shown explicitly that $c.l.(G) = 1$.

28.2. Small Values of q.

We look at $G = L_2(q)$ for $q \in \{5, 9, 13, 17, 25, 29\}$. Because $c.l.(G) = 1$ (see Remark 28.2) and because G can be generated by two elements (see for example [**Hup83**, II. 8.]), Corollary 23.3 yields that only $g_0 = [\chi, 1_G] \in \{0, 1\}$ must be considered.

The group $L_2(5)$ is isomorphic with $L_2(4) \cong A_5$, which has been treated already in Section 27.3; there is a unique exceptional character for $g_0 = 1$. Note that from the isomorphism $L_2(5) \cong L_2(4)$, it is clear that the generic exceptional character constructed in Section 27.2 for $L_2(q)$ with $q = 2^n$, $n \geq 2$, does not occur in the case $q \equiv 1 \pmod 4$. For, if we set $G = L_2(5) \cong L_2(4)$ and $C = P^G \cup Q^G$ (the elements of order $q = 5$ in G) then $\mathrm{Epi}_{(C)}(1, G) \neq \emptyset$, as we have seen in Table 21.

The group $L_2(9) \cong A_6$ turns out to have exactly the following four (RH)-characters that do not come from Riemann surfaces.

g	g_0	Nonzero $l_\chi(\sigma)$	χ
31	0	$l_\chi(3A) = 1, l_\chi(4A) = 2$	$\chi_3 + \chi_{4,5} + \chi_7$
31	0	$l_\chi(3B) = 1, l_\chi(4A) = 2$	$\chi_2 + \chi_{4,5} + \chi_7$
121	1	$l_\chi(3A) = 1$	$\chi_1 + \chi_2 + 2\chi_3 + 3\chi_{4,5} + 3\chi_6 + 3\chi_7$
121	1	$l_\chi(3B) = 1$	$\chi_1 + 2\chi_2 + \chi_3 + 3\chi_{4,5} + 3\chi_6 + 3\chi_7$

Each two characters of the same degree in this list are conjugate under automorphisms of $L_2(9)$.

For $q \in \{13, 17, 29\}$, every (RH)-character of $L_2(q)$ comes from a Riemann surface.

Finally, exactly the (RH)-characters χ for the following vectors l_χ of $L_2(25)$ do not come from Riemann surfaces.

g	g_0	$l_\chi(2A)$	$l_\chi(3A)$	$l_\chi(4A)$	$l_\chi(5A)$	$l_\chi(5B)$	$l_\chi(6A)$	$l_\chi(12A)$	$l_\chi(13A)$
1 821	0	0	0	0	0	1	2	0	0
2 016	0	0	0	0	0	2	0	1	0
1 821	0	0	0	0	1	0	2	0	0
1 691	0	0	0	0	1	1	1	0	0
1 561	0	0	0	0	1	2	0	0	0
2 016	0	0	0	0	2	0	0	1	0
1 561	0	0	0	0	2	1	0	0	0
1 496	0	0	0	1	0	1	1	0	0
1 366	0	0	0	1	0	2	0	0	0
1 496	0	0	0	1	1	0	1	0	0
1 366	0	0	0	1	2	0	0	0	0
1 171	0	0	0	2	0	1	0	0	0
1 171	0	0	0	2	1	0	0	0	0
1 301	0	0	1	0	0	0	2	0	0
1 041	0	0	1	0	1	1	0	0	0
976	0	0	1	1	0	0	1	0	0
651	0	0	1	2	0	0	0	0	0
651	0	1	0	0	0	0	2	0	0
391	0	1	0	0	1	1	0	0	0
326	0	1	0	1	0	0	1	0	0

29. Example: $G = S_n$

In this section, let $G = S_n$, the symmetric group on n points. Symmetric groups are ambivalent, see for example [**JK81**, Lemma 1.2.8]. In fact all characters of symmetric groups are rational by [**JK81**, Theorem 1.2.17].

29.1. Finiteness Result.

THEOREM 29.1. *Let $G = S_n$, with $n \geq 5$. There are only finitely many (RH)-characters of G that do not come from Riemann surfaces.*

Proof. The derived subgroup A_n of S_n, the alternating group on n points, has index 2 in S_n. For $n \geq 5$, A_n is simple. So it suffices to prove that the conditions of Theorem 23.15 are satisfied, and this is done in Lemma 29.2 below.   □

LEMMA 29.2. *Let G be an alternating group of degree at least 5. Then each element $\sigma \in G$ is the product of two elements that are conjugate in G to σ.*

Proof. In a *symmetric* group, two elements are conjugate if and only if they have the same cycle structure. The conjugacy class of an element in an *alternating* group is determined by its cycle structure if and only if its centralizer in the symmetric group contains odd permutations; this is the case, e.g., for every element of even order, since such an element contains a cycle of even length which centralizes and does not lie in an alternating group.

We show that each element σ in an alternating group $G = A_n$ of degree $n \geq 5$ is the product of two elements that have the same cycle structure as σ. For elements of odd order, we show additionally that the two factors can be chosen such that they are conjugate in A_n to σ.

First we solve the problem for σ a cycle of odd length or a product of two cycles of even length. Note that σ is a product of disjoint cycles where the number of even cycles is even. We are done if we can write each cycle or pair of cycles respectively as a product of two conjugates that move the same points and such that the conjugating permutations also move only these points. To see this, let $\sigma = \sigma_1 \sigma_2 \cdots \sigma_n$ be a representation of σ where the sets of moved points of σ_i and σ_j are disjoint for $i \neq j$. If $\sigma_i = \sigma_i^{\tau_i} \sigma_i^{\kappa_i}$ for all i then

$$
\begin{aligned}
\sigma &= \sigma_1^{\tau_1} \sigma_1^{\kappa_1} \cdots \sigma_n^{\tau_n} \sigma_n^{\kappa_n} \\
&= (\sigma_1^{\tau_1} \cdots \sigma_n^{\tau_n})(\sigma_1^{\kappa_1} \cdots \sigma_n^{\kappa_n}) \\
&= (\sigma_1 \cdots \sigma_n)^{\tau_1 \cdots \tau_n} (\sigma_1 \cdots \sigma_n)^{\kappa_1 \cdots \kappa_n} .
\end{aligned}
$$

The second of the above equalities holds because the relevant conjugates commute, and the third holds because the conjugators τ_i and κ_i respectively commute with each other and with σ_j for $i \neq j$.

So let us look at the following three special cases.

(a) If $\sigma = (1, 2, 3)$ then $\sigma = \sigma^2 \cdot \sigma^2$, and $\sigma^2 = \sigma^\tau$, with $\tau = (1, 2)(4, 5)$. Note that σ cannot be written as the product of two of its conjugates in A_3 and A_4.

(b) If $\sigma = (1, 2, \ldots, 2n + 1)$, for $n > 1$, then we can take a square root of σ of the same cycle type, and get $\sigma = \sigma^{n+1} \cdot \sigma^{n+1}$, with $\sigma^{n+1} = (1, n + 2, 2, n + 3, \ldots, n, 2n + 1, n + 1)$.

So we are done if σ and σ^{n+1} are conjugate in A_{2n+1}. Otherwise (see Remark 29.3 below) we set $\tau = (1, 2)(n + 2, n + 3)$, and write $\sigma = (\sigma^{n+1} \tau) \cdot (\tau \sigma^{n+1})$. Then

$$
\sigma^{n+1} \tau = (1, n + 3, 3, n + 4, 4, \ldots, 2n + 1, n + 1, 2, n + 2)
$$

and

$$
\tau \sigma^{n+1} = (1, n + 3, 2, n + 2, 3, n + 4, 4, \ldots, 2n + 1, n + 1)
$$

are both conjugate to σ^{n+1} in S_{2n+1}; for example, conjugating σ^{n+1} with $(2, 3, \ldots, 2n+1)$ and $(n+2, n+3)$ yields $\sigma^{n+1} \tau$ and $\tau \sigma^{n+1}$, respectively. These conjugating permutations lie in $S_{2n+1} \setminus A_{2n+1}$, and since σ and σ^{n+1} are also conjugate in $S_{2n+1} \setminus A_{2n+1}$, we get that $\sigma^{n+1} \tau$ and $\tau \sigma^{n+1}$ are conjugate to σ in A_{2n+1}.

(c) If $\sigma = (1, 2, \ldots, 2n)(2n+1, 2n+2, \ldots, 2n+2m)$, with $n \geq 1$ and $m \geq 1$, then we set $\tau = (2n - 1, 2n, 2n + 1) \cdot (2n + 2, 2n + 3, \ldots, 2n + 2m)^{-1}$. So $\tau \in A_{2n+1}$,

$$\sigma^\tau = (1, 2, \ldots, 2n - 2, 2n, 2n + 1)$$
$$(2n - 1, 2n + 2m, 2n + 2, 2n + 3, \ldots, 2n + 2m - 1)$$

and

$$\sigma^\tau \sigma = (1, 3, 5, \ldots, 2n + 1, 2, 4, 6, \ldots, 2n - 2)$$
$$(2n, 2n + 2, \ldots, 2n + 2m, 2n + 3, 2n + 5, \ldots, 2n + 2m - 1),$$

hence we have written the right hand side, which is a permutation that consists of two disjoint cycles of lengths $2n$ and $2m$, as a product of the two elements σ^τ and σ with the same cycle structure.

In the general case that σ is the product of disjoint cycles, we see that only case (a) causes problems because the conjugating permutation moves points different from those moved by the 3-cycle. So we have to change our strategy if σ contains an odd number of 3-cycles and has fewer than two fixed points.

If σ contains at least two 3-cycles (i_1, i_2, i_3) and (j_1, j_2, j_3) then a conjugating permutation may be multiplied by $(i_1, j_1)(i_2, j_2)(i_3, j_3)$, which centralizes σ and is odd. If σ contains a *single* 3-cycle (i_1, i_2, i_3) then σ must have a cycle of length different from 3, so one of the cases (b) and (c) occurs, and we have seen above that we may choose an *odd* conjugating permutation in each of these cases, which together with $\tau = (i_1, i_2)$ gives an even permutation.   □

REMARK 29.3. One can show that the two possibilities in case (b) of the above proof do both occur. More precisely, for odd n, the cycles $\sigma = (1, 2, \ldots, n)$ and σ^2 are conjugate in A_n if and only if $n \equiv \pm 1 \pmod 8$.

COROLLARY 29.4. *Lemma 29.2 implies that c.l.*$(A_n) = c.l.(S_n) = 1$ *for* $n \geq 5$, *see Remark 27.3. A direct proof of this fact can be found in* [Itō51].

29.2. Small Values of n. We classify the characters of S_n, for $2 \leq n \leq 5$, that come from Riemann surfaces.

S_2 is cyclic of order 2, it has been shown in Section 25 that every (RH)-character of S_2 comes from a Riemann surface. The same has been shown for $S_3 \cong D_6$ in Section 21. For S_4, there is one infinite series of exceptions, see Example 24.6. Every (RH)-character of S_5 comes from a Riemann surface.

29.3. Ambivalent Alternating Groups. Lemma 29.2 does *not* prove that all except finitely many (RH)-characters of the alternating group A_n, $n \geq 5$, come from Riemann surfaces. More precisely, the lemma proves this only for the case that A_n is ambivalent. By [JK81, Lemma 1.2.12], the alternating group A_n is ambivalent if and only if $n \in \{0, 1, 2, 5, 6, 10, 14\}$. The groups A_n with $n \leq 2$ are trivial, and the cases $A_5 \cong L_2(4)$ and $A_6 \cong L_2(9)$ have been studied in Sections 27.3 and 28.2.

The groups A_{10} and A_{14} seem to be out of reach for a full classification with the help of Algorithm 24.4; in Step 3 of this algorithm, looping over all vectors l with $0 \leq l(H) \leq q(H) + 1$ would mean looping over $7\,527\,168\,000\,000\,000$ vectors for A_{10}.

CHAPTER 7

Classification for Fixed Group: Nonreal Irrationalities

In this chapter, we generalize the results of Chapter 6 to the situation of not necessarily ambivalent groups G, that is, the character table of G may contain nonreal entries.

It will turn out that the number theoretic property (LI) that is studied in Appendix C (see Definition C.1) is useful for practical purposes.

30. Necessary Conditions on χ

The presentation of the material follows Section 22. There we used the statements of Section 10, now we use their refinements as stated in Section 11.

30.1. The Condition $(\widetilde{\text{E}})$. We have introduced the condition (E) as an abstraction of the Lefschetz Fixed Point Formula (see Corollary 12.3). Now we define an analogous condition that describes the Eichler Trace Formula (see Theorem 12.1).

DEFINITION 30.1 (cf. Definition 22.1). We say that the number $\alpha \in \mathbb{Q}(\zeta_m)$ satisfies the Eichler Trace Formula for m if α can be written as $\alpha = 1 + \sum_{u \in I(m)} f_u \frac{\zeta_m^u}{1 - \zeta_m^u}$ with nonnegative integers f_u. We say that the G-character χ satisfies the condition $(\widetilde{\text{E}})$ if $\chi(1) \geq 2$ and for all $\sigma \in G^\times$, $\chi(\sigma)$ satisfies the Eichler Trace Formula for $|\sigma|$.

EXAMPLE 30.2. A *real* algebraic integer $\alpha \in \mathbb{Q}(\zeta_m)$ satisfies the Eichler Trace Formula for m if and only if $\alpha \in \mathbb{Z}$ with $\alpha \leq 1$. That is, for one direction we have

$$2\alpha = \alpha + \overline{\alpha} = 2 - \sum_{u \in I(m)} f_u$$

by Lemma 12.2, so $2\alpha \in \mathbb{Z}$ with $\alpha \leq 1$, and $2\alpha \in \mathbb{Z}$ implies $\alpha \in \mathbb{Z}$ because α was assumed to be an algebraic integer. For the other direction, we may set

$$f_1 = f_{m-1} = 1 - \alpha \quad \text{and} \quad f_u = 0 \quad \text{for} \quad u \in I(m) \setminus \{1, m-1\},$$

138

then all f_u are nonnegative integers, and we get

$$1 + \sum_{u \in I(m)} f_u \frac{\zeta_m^u}{1 - \zeta_m^u} = 1 + (1 - \alpha) \left(\frac{\zeta_m}{1 - \zeta_m} + \frac{\zeta_m^{-1}}{1 - \zeta_m^{-1}} \right) = \alpha.$$

The assumption that α is an algebraic integer can be dropped if m has the property (LI), since in this case a real number α that satisfies the Eichler Trace Formula can be expressed uniquely as

$$\alpha = \left(1 - \sum_{u \in I} f_{m-u} \right) + \sum_{u \in I} (f_u - f_{m-u}) \frac{\zeta_m^u}{1 - \zeta_m^u}$$

$$= \left(1 - \sum_{u \in I} \frac{f_u + f_{m-u}}{2} \right) + \sum_{u \in I} (f_u - f_{m-u}) \left(\frac{\zeta_m^u}{1 - \zeta_m^u} + \frac{1}{2} \right)$$

for any complementary subset (see Definition C.1) I of $I(m)$. All summands except the first are purely imaginary by Lemma 12.2, so $f_u = f_{m-u}$ holds for all $u \in I$, and $f_u + f_{m-u}$ is even.

We are only interested in character values, which are always algebraic integers. (Note that the real number $1 + \frac{\zeta_2}{1 - \zeta_2} = \frac{1}{2}$ satisfies the Eichler Trace Formula for $m = 2$ but is not integral.) ◇

REMARK 30.3. A character χ of G that comes from a Riemann surface satisfies the condition $(\widetilde{E})$ by Theorem 12.1, and the condition $(\widetilde{E})$ implies the condition (E) (see Definition 22.1) because

$$\chi(\sigma) + \overline{\chi(\sigma)} = 2 + \sum_{u \in I(m)} f_u \left(\frac{\zeta_m^u}{1 - \zeta_m^u} + \frac{\zeta_m^{-u}}{1 - \zeta_m^{-u}} \right) = 2 - \sum_{u \in I(m)} f_u$$

for $\sigma \in G^\times$ by Lemma 12.2, so $\chi(\sigma) + \overline{\chi(\sigma)}$ is an integer that is smaller than or equal to 2.

Moreover, Example 30.2 shows that a *real* character χ satisfies the condition $(\widetilde{E})$ if and only if it satisfies the condition (E) and additionally $\chi(\sigma) \leq 1$ for all $\sigma \in G^\times$ (cf. Remark 22.2). In particular, for the treatment of ambivalent groups, the condition $(\widetilde{E})$ does not provide a new criterion.

In order to decide whether a given number $\alpha \in \mathbb{Q}(\zeta_m)$ satisfies the Eichler Trace Formula for m, we must consider only finitely many possible coefficients f_u, since they are nonnegative integers with $\sum_{u \in I(m)} f_u = 2 - (\alpha + \overline{\alpha})$. For a complementary subset I of $I(m)$, we can write α in the form $\alpha = a_0 + \sum_{u \in I} a_u \frac{\zeta_m^u}{1 - \zeta_m^u}$ by Lemma 12.2, and the coefficients a_u are unique if m has the property (LI).

LEMMA 30.4. *Let m be a positive integer, I a complementary subset of $I(m)$, and*

$$\alpha = a_0 + \sum_{u \in I} a_u \frac{\zeta_m^u}{1 - \zeta_m^u},$$

with rational numbers a_0 and a_u, for $u \in I$. If

$$(*) \qquad a_0 \quad \text{and} \quad a_u \in \mathbb{Z} \quad \text{for} \quad u \in I \quad \text{are such that} \quad a_0 - \sum_{\substack{u \in I \\ a_u < 0}} a_u \leq 1$$

then α satisfies the Eichler Trace Formula for m. If m has the property (LI) *then the converse also holds.*

Proof. If $(*)$ holds then $\delta = 1 - a_0 + \sum_{u \in I, a_u < 0} a_u$ is a nonnegative integer. We choose $u_0 \in I$, and set

$$f_{u_0} = \delta + \max\{0, a_{u_0}\}, f_u = \max\{0, a_u\} \quad \text{for} \quad u \in I \setminus \{u_0\},$$

and $f_{m-u} = f_u - a_u$ for $u \in I$. Then the f_u are nonnegative integers, $f_u - f_{m-u} = a_u$ holds, and Lemma 12.2 yields

$$\begin{aligned}
\alpha &= a_0 + \sum_{u \in I} (f_u - f_{m-u}) \frac{\zeta_m^u}{1 - \zeta_m^u} \\
&= \left(a_0 + \sum_{u \in I} f_{m-u} \right) + \sum_{u \in I(m)} f_u \frac{\zeta_m^u}{1 - \zeta_m^u}.
\end{aligned}$$

The first summand is equal to

$$a_0 + \sum_{u \in I} (f_u - a_u) = a_0 + \delta - \sum_{\substack{u \in I \\ a_u < 0}} a_u = 1,$$

so α satisfies the Eichler Trace Formula for m.

Conversely, if α satisfies the Eichler Trace Formula for m then

$$\alpha = 1 + \sum_{u \in I(m)} f_u \frac{\zeta_m^u}{1 - \zeta_m^u} = \left(1 - \sum_{u \in I} f_{m-u} \right) + \sum_{u \in I} (f_u - f_{m-u}) \frac{\zeta_m^u}{1 - \zeta_m^u}$$

by Lemma 12.2, with nonnegative integers f_u. We assumed that m has the property (LI), so the coefficients

$$a_0 = \left(1 - \sum_{u \in I} f_{m-u} \right) \quad \text{and} \quad a_u = (f_u - f_{m-u})$$

are uniquely determined integers, and

$$a_0 - \sum_{\substack{u \in I \\ a_u < 0}} a_u = 1 - \sum_{\substack{u \in I \\ a_u < 0}} f_u - \sum_{\substack{u \in I \\ a_u \geq 0}} f_{m-u} \leq 1.$$

$$\square$$

If m has the property (LI) then we can show that the inequality $(*)$ in Lemma 30.4 is independent of the choice of the complementary subset of $I(m)$.

LEMMA 30.5. *Let m be a positive integer with the property* (LI), *and I and J two complementary subsets of $I(m)$. If*

$$a_0 + \sum_{u \in I} a_u \frac{\zeta_m^u}{1 - \zeta_m^u} = b_0 + \sum_{u \in J} b_u \frac{\zeta_m^u}{1 - \zeta_m^u}$$

then

$$a_0 - \sum_{\substack{u \in I \\ a_u < 0}} a_u = b_0 - \sum_{\substack{u \in J \\ b_u < 0}} b_u.$$

Proof. We are done if we can show the claim for the case that J is obtained from I by replacing a single element. So let $u_0 \in I$ with $I \backslash \{u_0\} = J \backslash \{m - u_0\}$. Then

$$
\begin{aligned}
a_0 + \sum_{u \in I} a_u \frac{\zeta_m^u}{1 - \zeta_m^u} &= a_0 + a_{u_0} \frac{\zeta_m^{u_0}}{1 - \zeta_m^{u_0}} + \sum_{\substack{u \in I \\ u \neq u_0}} a_u \frac{\zeta_m^u}{1 - \zeta_m^u} \\
&= (a_0 - a_{u_0}) - a_{u_0} \frac{\zeta_m^{m - u_0}}{1 - \zeta_m^{m - u_0}} + \sum_{\substack{u \in J \\ u \neq m - u_0}} a_u \frac{\zeta_m^u}{1 - \zeta_m^u},
\end{aligned}
$$

so $b_0 = a_0 - a_{u_0}$, $b_u = a_u$ for $u \in I \setminus \{u_0\}$, and $b_{m - u_0} = -a_{u_0}$ holds by the property (LI). Hence

$$
\begin{aligned}
b_0 - \sum_{\substack{u \in J \\ b_u < 0}} b_u &= a_0 - (a_{u_0} + \min\{0, b_{m - u_0}\}) - \sum_{\substack{u \in I \backslash \{u_0\} \\ a_u < 0}} a_u \\
&= a_0 - \min\{0, a_{u_0}\} - \sum_{\substack{u \in I \backslash \{u_0\} \\ a_u < 0}} a_u \\
&= a_0 - \sum_{\substack{u \in I \\ a_u < 0}} a_u.
\end{aligned}
$$

$$\square$$

REMARK 30.6. The number α in Lemma 30.4 is usually given as a linear combination of the values of irreducible characters. Because

$$a_0 - \sum_{\substack{u \in I \\ a_u < 0}} a_u = \min \left\{ a_0 - \sum_{\substack{u \in J \\ a_u < 0}} a_u \,\middle|\, J \subseteq I \right\},$$

the inequality in line ($*$) of Lemma 30.4 holds if and only if $a_0 - \sum_{u \in J} a_u \leq 1$ for all subsets $J \subseteq I$. So the question whether α satisfies the Eichler Trace Formula for m can be translated into $2^{\varphi(m)/2}$ inequalities in terms of the coefficients of this linear combination.

REMARK 30.7. If we are interested in the question whether a given character χ comes from a Riemann surface then we can postulate a stronger condition than $(\widetilde{E})$, which yields a smaller number of inequalities. That is, we may identify, for $h \in G^\times$ of order m, the conjugation action of $N_G(\langle h \rangle)/C_G(h)$ on the set $\{h^u \mid u \in I(m)\}$ and the action of the corresponding group of prime residues modulo m on the set $I(m)$; then Lemma 11.3 yields that we need to consider only those expressions $\chi(h) = 1 + \sum_{u \in I(m)} f_u \frac{\zeta_m^u}{1 - \zeta_m^u}$ where f_u as a function of u is constant on orbits of $N_G(\langle h \rangle)/C_G(h)$.

30.2. Example: $G = C_3$. We show that a character of the cyclic group $G = \langle a \mid a^3 \rangle$ of order 3 comes from a Riemann surface if and only if it satisfies the condition $(\widetilde{E})$. (This will be generalized in Section 32.) The character table of G is shown in Table 23.

	1	a	a^2
1_G	1	1	1
χ_2	1	ζ_3	ζ_3^2
χ_3	1	ζ_3^2	ζ_3

TABLE 23. Irreducible Characters of C_3

Clearly every character of G of degree at least 2 satisfies the condition (E), and $\chi = a_1 1_G + a_2 \chi_2 + a_3 \chi_3$ satisfies the condition (RH) if and only if

$$l_\chi(a) = r_\chi^*(a) = r_\chi(a) = 2 - (\chi(a) + \overline{\chi(a)}) = 2 - 2a_1 + a_2 + a_3 \geq 0,$$

because $\zeta_3 + \zeta_3^2 = -1$.

Now we look at the condition $(\widetilde{E})$. We have $\frac{\zeta_3}{1 - \zeta_3} = \frac{1}{3}(\zeta_3 - 1)$, and since 3 has the property (LI), $\chi(a)$ can be written uniquely as a linear combination of 1 and $\frac{\zeta_3}{1 - \zeta_3}$, that is, $\chi(a) = (a_1 + a_2 - 2a_3) + (3a_2 - 3a_3)\frac{\zeta_3}{1 - \zeta_3}$. By Lemma 30.4 and Remark 30.6, χ is a character that satisfies the condition $(\widetilde{E})$ if and only if

$$
\begin{aligned}
a_1, \quad a_2, \quad a_3 &\geq 0, \\
a_1 + a_2 - 2a_3 &\leq 1, \\
a_1 - 2a_2 + a_3 &\leq 1, \quad \text{and} \\
a_1 + a_2 + a_3 &\geq 2 \,.
\end{aligned}
$$

Figure 3 shows the integral solutions of these inequalities in the (a_2, a_3)–plane for $a_1 \in \{0, 1, 2\}$, similarly to Figure 1 on page 94. Note that the condition (RH) is implied by the condition $(\widetilde{E})$.

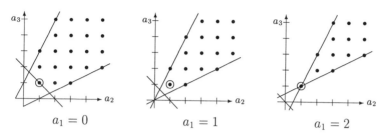

$$a_1 = 0 \qquad\qquad a_1 = 1 \qquad\qquad a_1 = 2$$

FIGURE 3. Realizable Characters for C_3, with $a_1 \leq 2$.

Finally, we show that each character of G that satisfies the condition $(\widetilde{E})$ does in fact come from a Riemann surface. The argumentation is analogous to that in Section 21.

The circled points in Figure 3 correspond to the characters $\mathrm{Tr}(\Phi_k)$, $0 \leq k \leq 2$, where the surface kernel epimorphisms $\Phi_k \colon \Gamma(k; m_1, m_2, \dots, m_r)$ with $r = 4 - 2k$ and $m_i = 3$ for $1 \leq i \leq r$ map $r/2$ each of the elliptic generators to a and a^2. To see this, we set $X = \mathcal{U}/\ker(\Phi_k)$, and compute

$$\mathrm{Tr}(\Phi_k)(1) = g(X) = 1 + (k - 1) \cdot 3 + (3/2) \cdot r \cdot (1 - 1/3) = k + 2$$

and $|\mathrm{Fix}_{X,1}(a)| = |\mathrm{Fix}_{X,2}(a)| = r/2 = 2 - k$, so

$$\mathrm{Tr}(\Phi_k)(a) = 1 + (2 - k) \cdot \frac{\zeta_3}{1 - \zeta_3} + (2 - k) \cdot \frac{\zeta_3^2}{1 - \zeta_3^2} = 1 - (2 - k) = k - 1,$$

and thus $\mathrm{Tr}(\Phi_k) = k \cdot 1_G + \chi_2 + \chi_3$.

Adding two new elliptic generators and mapping one each to a and a^2 increases the character degree by $(3/2) \cdot 2 \cdot (1 - 1/3) = 2$, increases both $|\mathrm{Fix}_{X,1}(a)|$ and $|\mathrm{Fix}_{X,2}(a)|$ by 1, and hence decreases the character value on a by 1. So this transformation maps χ to $\chi + \chi_2 + \chi_3$. In terms of Figure 3, we have established all characters corresponding to the interesting points on the main diagonals.

To get surface kernel epimorphisms for the other characters with $a_1 \leq 2$, we consider the transformation that replaces one elliptic generator that is mapped to a^2 by two elliptic generators that are both mapped to a. It increases the character degree by 1, increases $|\mathrm{Fix}_{X,1}(a)|$ by 2, decreases $|\mathrm{Fix}_{X,2}(a)|$ by 1, and thus changes $\chi(a)$ to $\chi(a) + 2\frac{\zeta_3}{1 - \zeta_3} - \frac{\zeta_3^2}{1 - \zeta_3^2} = \chi(a) + \zeta_3$. So this transformation maps χ to $\chi + \chi_2$. Analogously, replacing an image a by two images a^2 maps χ to $\chi + \chi_3$.

Finally, if $a_1 > 2$, all interesting characters are described exactly by the two inequalities of condition $(\widetilde{E})$, $a_1 + a_2 - 2a_3 \leq 1$ and $a_1 - 2a_2 + a_3 \leq 1$. For each solution, both a_2 and a_3 are strictly positive, and $\chi - \rho_G$ also satisfies the system of inequalities. By induction on a_1, we may take a surface kernel epimorphism for $\chi - \rho_G$, and construct one for χ by increasing the orbit genus by 1.

30.3. The Condition $(\widetilde{RH})$. The final insight of Section 22.2 was that if a G-class function χ satisfies the condition (E) then $\chi + \overline{\chi}$ can be written as a linear combination of certain induced class functions, with nonnegative integral coefficients if and only if the condition (RH) is satisfied. Now we try to get an analogous description for the condition $(\widetilde{E})$ instead of (E).

DEFINITION 30.8 (cf. Definition 22.1). Let G be a finite group, and χ a class function of G that satisfies the condition $(\widetilde{E})$. If nonnegative integers $r_{\chi,u}(h)$ are given such that

$$\chi(h) = 1 + \sum_{u \in I(|h|)} r_{\chi,u}(h) \frac{\zeta_m^u}{1 - \zeta_m^u}$$

holds for all $h \in G^\times$ then the values

$$r_{\chi,u}^*(h) = r_{\chi,u}(h) - \sum_{k \in cy(G,h)} \sum_{\substack{v \in I(|k|) \\ v \equiv u \pmod{|h|}}} r_{\chi,v}^*(k)$$

and

$$l_{\chi,u}(h) = r_{\chi,u}^*(h)/[C_G(h):\langle h \rangle]$$

are well-defined. In this case, we call $(l_{\chi,u}(h))_{h \in G^\times, u \in I(|h|)}$ an *admissible system for* χ if the underlying map $r_{\chi,u}$ on $G^\times$ is constant on conjugacy classes of G and $r_{\chi,u}(h^u) = r_{\chi,1}(h)$ holds for all $h \in G^\times$ and $u \in I(|h|)$.

We say that χ is an $(\widetilde{RH})$-class function (or that χ satisfies the condition $(\widetilde{RH})$) if and only if χ satisfies the condition $(\widetilde{E})$ and there is an admissible system $(l_{\chi,u}(h))_{h \in G^\times, u \in I(|h|)}$ such that $l_{\chi,u}(h)$ is a nonnegative integer for all $h \in G^\times$ and $u \in I(|h|)$.

A character of G that comes from a Riemann surface X satisfies the condition $(\widetilde{RH})$ with $r_{\chi,u}(h) = |\text{Fix}_{X,u}(h)|$ (see Theorem 12.1), $r_{\chi,u}^*(h) = |\text{Fix}_{X,u}^G(h)|$ (see Lemma 11.2), and $l_{\chi,u}(h) = |\pi(\text{Fix}_{X,u}^G(h))|$ (see Lemma 11.4), where π denotes the natural projection $X \to X/G$.

It should be emphasized that in spite of the notation we introduced above, a class function χ does *not* in general determine the values $r_{\chi,u}(h)$, $r_{\chi,u}^*(h)$, and $l_{\chi,u}(h)$ uniquely, see Example 30.20 below. This notation was chosen parallel to the one in Definition 22.1.

Note that conversely, the numbers $(l_{\chi,u}(h))_{h\in G^\times, u\in I(|h|)}$ determine the class function χ by definition. In other words, if $(l_{\chi,u}(h))_{h\in G^\times, u\in I(|h|)}$ is an admissible system for some class function then it is in fact an admissible system for χ.

REMARK 30.9. In order to define an admissible system for a given character χ, it is sufficient either to choose all $r_{\chi,u}(h)$, $u \in I(|h|)$, for representatives h of nontrivial cyclic subgroups of G or alternatively to choose $r_{\chi,1}(h)$ for representatives h of G-conjugacy classes in $G^\times$.

By Lemma 11.5, the values $l_{\chi,1}(h)$ can be interpreted as the number of elliptic generators of $\Gamma(\chi)$ that are mapped to a G-conjugate of h under any surface kernel epimorphism Φ with $\mathrm{Tr}(\Phi) = \chi$. This is a refinement of the condition (RH) which controls the distribution of the images of elliptic generators to the conjugacy classes of cyclic subgroups in G. Note that the data of the conditions (RH) and $\widetilde{(\mathrm{RH})}$ are compatible in the following sense. For $h \in G^\times$ of order m,

$$r_\chi(h) = 2 - (\chi(h) + \overline{\chi}(h)) = - \sum_{u\in I(m)} r_{\chi,u}\left(\frac{\zeta_m^u}{1-\zeta_m^u} + \frac{\zeta_m^{-u}}{1-\zeta_m^{-u}} \right) = \sum_{u\in I(m)} r_{\chi,u},$$

and by the recursive definition of $r_{\chi,u}^*(h)$, we have

$$
\begin{aligned}
\sum_{u\in I(m)} r_{\chi,u}^*(h) &= \sum_{u\in I(m)} \left(r_{\chi,u}(h) - \sum_{k\in cy(G,h)} \sum_{\substack{v\in I(|k|) \\ v\equiv u \ (\mathrm{mod}\ m)}} r_{\chi,v}^*(k) \right) \\
&= r_\chi(h) - \sum_{k\in cy(G,h)} \sum_{v\in I(|k|)} r_{\chi,v}^*(k) \\
&= r_\chi(h) - \sum_{k\in cy(G,h)} r_\chi^*(k) \\
&= r_\chi^*(h).
\end{aligned}
$$

As a consequence,

$$[N_G(\langle h\rangle):C_G(h)] \cdot l_\chi(h) = \sum_{u\in I(m)} l_{\chi,u}(h).$$

By Remark 30.7, we can write $l_\chi(h)$ as a sum of $l_{\chi,u}(h)$ where u runs over a set of orbits of $N_G(\langle h\rangle)/C_G(h)$ on $I(m)$. In particular, the condition $\widetilde{(\mathrm{RH})}$ implies the condition (RH).

Under the assumption of the property (LI), we can describe the ambiguities in the choice of admissible systems for a class function.

LEMMA 30.10. *Let G be a finite group such that all element orders larger than 2 of G have the property (LI), χ a G-class function that satisfies the* *condition* $(\widetilde{E})$, *and* $(l_{\chi,u}(h))_{h\in G^\times, u\in I(|h|)}$ *an admissible system for* χ. *Then for each* $h \in G^\times$, *the differences* $l_{\chi,u}(h) - l_{\chi,u}(h^{-1})$ *are determined by* χ.

Proof. For $h \in G$ of order 2, there is nothing to show, so let $h \in G^\times$ with $|h| = m > 2$, and I be a complementary subset of $I(m)$. We have

$$\chi(h) = 1 + \sum_{u\in I(m)} r_{\chi,u}(h)\frac{\zeta_m^u}{1 - \zeta_m^u}$$

$$= \left(1 - \sum_{u\in I} r_{\chi,m-u}(h)\right) + \sum_{u\in I}\left(r_{\chi,u}(h) - r_{\chi,m-u}(h)\right)\frac{\zeta_m^u}{1 - \zeta_m^u},$$

and since m has the property (LI), the coefficients $r_{\chi,u}(h) - r_{\chi,m-u}(h) = r_{\chi,u}(h) - r_{\chi,u}(h^{-1})$, for $u \in I$, can be computed from $\chi(h)$.

By the recursive definition of $r^*_{\chi,u}(h)$, also $r^*_{\chi,u}(h) - r^*_{\chi,m-u}(h)$

$$= \left(r_{\chi,u}(h) - r_{\chi,m-u}(h)\right)$$

$$- \sum_{k\in cy(G,h)}\left(\sum_{\substack{v\in I(|k|)\\v\equiv u \,(\mathrm{mod}\, m)}} r^*_{\chi,v}(k) - \sum_{\substack{v\in I(|k|)\\v\equiv m-u \,(\mathrm{mod}\, m)}} r^*_{\chi,v}(k)\right)$$

$$= \left(r_{\chi,u}(h) - r_{\chi,m-u}(h)\right) - \sum_{k\in cy(G,h)}\sum_{\substack{v\in I(|k|)\\v\equiv u \,(\mathrm{mod}\, m)}}\left(r^*_{\chi,v}(k) - r^*_{\chi,|k|-v}(k)\right)$$

and thus $l_{\chi,u}(h) - l_{\chi,m-u}(h) = l_{\chi,u}(h) - l_{\chi,u}(h^{-1})$ can be recovered from χ. $\square$

DEFINITION 30.11. Let G be a group, and $h \in G^\times$. Define $\varepsilon_h \in \mathrm{Irr}(\langle h\rangle)$ by $\varepsilon_h(h) = \zeta_{|h|}$, and the G-class function $\widetilde{\psi}_{G,h}$ by

$$\widetilde{\psi}_{G,h} = \sum_{i=1}^{|h|-1} \frac{i}{|h|}\left(\varepsilon_h^{-i}\right)^G.$$

Furthermore, set $\widetilde{\psi}_{G,1} = \rho_G$.

THEOREM 30.12 (cf. Corollary 22.5). *Let* χ *be a* G-*class function that satisfies the condition* $(\widetilde{E})$, *and* $(l_{\chi,u}(h))_{h\in G^\times, u\in I(|h|)}$ *be an admissible system for* χ. *Set* $g_0 = [\chi, 1_G]$. *Then*

$$\chi = 1_G + (g_0 - 1)\cdot \rho_G + \sum_{h\in G^\times/\sim_G} l_{\chi,1}(h)\cdot\widetilde{\psi}_{G,h}.$$

Proof. Take $h \in G^\times$ of order m. First we reformulate the relation between $r_{\chi,u}(h)$ and $r^*_{\chi,u}(h)$. That is, for $k \in cy(G,h)$ and $v \in I(|k|)$ with $v \equiv 1$ (mod m), let v' denote the unique element of $I(|k|)$ such that $v\cdot v' \equiv 1$

(mod $|k|$) holds. Then $v' \equiv 1$ (mod m) and thus $(k^{v'})^{|k|/m} = h^{v'} = h$. Now $r^*_{\chi,v}(k) = r^*_{\chi,1}(k^{v'})$ implies that

$$\sum_{\substack{v \in I(|k|) \\ v \equiv 1 \,(\mathrm{mod}\ m)}} r^*_{\chi,v}(k) = \sum_{\substack{v \in I(|k|) \\ v \equiv 1 \,(\mathrm{mod}\ m)}} r^*_{\chi,1}(k^v)$$

is the sum over $r^*_{\chi,1}(\sigma)$ for all $\sigma \in \langle k \rangle$ with the property that $\sigma^{|k|/m} = h$ holds. Therefore,

$$
\begin{aligned}
r_{\chi,1}(h) &= r^*_{\chi,1}(h) + \sum_{k \in cy(G,h)} \sum_{\substack{v \in I(|k|) \\ v \equiv 1 \,(\mathrm{mod}\ m)}} r^*_{\chi,v}(k) \\
&= \sum_{\substack{H \in CY(G) \\ h \in H}} \sum_{\substack{k \in H \\ \langle k \rangle = H \\ k^{|H|/m} = h}} r^*_{\chi,1}(k) \\
&= \sum_{\sigma \in G^\times / \sim_G} \sum_{\substack{H \in CY(G) \\ h \in H}} \sum_{\substack{k \in H \\ \langle k \rangle = H \\ k^{|H|/m} = h \\ k \in \sigma^G}} r^*_{\chi,1}(\sigma) \\
&= \sum_{\sigma \in G^\times / \sim_G} \sum_{\substack{k \in \sigma^G \\ k^{|k|/m} = h}} r^*_{\chi,1}(\sigma) \\
&= \sum_{\sigma \in G^\times / \sim_G} |\{k \mid k \in \sigma^G, k^{|k|/m} = h\}| \cdot r^*_{\chi,1}(\sigma).
\end{aligned}
$$

Note that we need this only for $u = 1$, we would not necessarily have $u \in I(|\sigma|)$ for general $u \in I(m)$ and $\sigma \in G^\times$.

Next we compute the cardinality of the set on the right hand side. We set $\delta(\sigma, h) = 1$ if σ is G-conjugate to an element $k \in G$ with the property that $k^{|k|/m} = h$, and $\delta(\sigma, h) = 0$ otherwise. Equivalently, $\delta(\sigma, h) = 1$ if and only if $\sigma^{|\sigma|/m} \sim_G h$ holds. With this notation, we claim that

$$|\{k \mid k \in \sigma^G, k^{|k|/m} = h\}| = \frac{|C_G(h)|}{|C_G(\sigma)|} \cdot \delta(\sigma, h).$$

To see this, observe that if $\delta(\sigma, h) = 0$ then no G-conjugate k of σ satisfies $k^{|k|/m} = h$, so the left hand side is zero. In the remaining case, h has a root in the class σ^G, and since each element in h^G has the same number of roots in σ^G, this number is equal to $|\sigma^G|/|h^G| = |C_G(h)|/|C_G(\sigma)|$.

As above, let u' denote the unique element of $I(m)$ with $u \cdot u' \equiv 1$ (mod m). Then we have

$$r_{\chi,u}(h) = r_{\chi,1}(h^{u'}) = \sum_{\sigma \in G^\times / \sim_G} \frac{|C_G(h)|}{|C_G(\sigma)|} \cdot r^*_{\chi,1}(\sigma) \cdot \delta(\sigma, h^{u'})$$

and thus

$$
\sum_{u\in I(m)} r_{\chi,u}(h)\frac{\zeta_m^u}{1-\zeta_m^u} = \sum_{\sigma\in G^\times/\sim_G} \frac{|C_G(h)|}{|C_G(\sigma)|}\cdot r_{\chi,1}^*(\sigma)\sum_{u\in I(m)}\delta(\sigma,h^{u'})\cdot\frac{\zeta_m^u}{1-\zeta_m^u}
$$

$$
= \sum_{\sigma\in G^\times/\sim_G} l_{\chi,1}(\sigma)\cdot\frac{|C_G(h)|}{|\sigma|}\sum_{\substack{u\in I(m)\\ \sigma^{|\sigma|/m}\in(h^{u'})^G}}\frac{\zeta_m^u}{1-\zeta_m^u}
$$

$$
= \sum_{\sigma\in G^\times/\sim_G} l_{\chi,1}(\sigma)\sum_{i=1}^{m-1}\frac{i}{|\sigma|}\cdot\frac{|C_G(h)|}{m}\sum_{\substack{u\in I(m)\\ \sigma^{u|\sigma|/m}\in h^G}}\zeta_m^{-ui},
$$

where we applied Lemma C.5 in the last step.

For fixed σ, we now set $d=|\sigma|/m$. In the equality

$$
\sum_{i=1}^{|\sigma|-1} i\cdot\zeta_m^{vi} = \sum_{c=0}^{d-1}\sum_{j=0}^{m-1}(j+cm)\cdot\zeta_m^{v(j+cm)} = d\sum_{j=0}^{m-1}j\cdot\zeta_m^{vj} + \sum_{c=0}^{d-1}cm\sum_{j=0}^{m-1}\zeta_m^{vj},
$$

with arbitrary $v\in I(m)$, the second sum on the far right hand side vanishes, which yields

$$
\sum_{u\in I(m)} r_{\chi,u}(h)\frac{\zeta_m^u}{1-\zeta_m^u} = \sum_{\sigma\in G^\times/\sim_G} l_{\chi,1}(\sigma)\sum_{i=1}^{|\sigma|-1}\frac{i}{|\sigma|}\cdot\frac{|C_G(h)|}{|\sigma|}\sum_{\substack{u\in I(m)\\ \sigma^{u|\sigma|/m}\in h^G}}\zeta_m^{-ui}.
$$

Finally, we note that (see page 23)

$$
\left(\varepsilon_\sigma^i\right)^G(h) = \frac{|C_G(h)|}{|\sigma|}\cdot\sum_{\substack{\tau\in\langle\sigma\rangle\\ \tau\in h^G}}\varepsilon_\sigma^i(\tau) = \frac{|C_G(h)|}{|\sigma|}\sum_{\substack{u\in I(m)\\ \sigma^{u|\sigma|/m}\in h^G}}\varepsilon_\sigma^i(\sigma^{u|\sigma|/m})
$$

and that $\varepsilon_\sigma^i(\sigma^{u|\sigma|/m}) = \zeta_{|\sigma|}^{ui|\sigma|/m} = \zeta_m^{ui}$.

Thus the claim is shown for all values $h\in G^\times$. For the degree, we use the same argument as in the motivation of Corollary 22.5, namely, the difference of the two class functions in question is a multiple of the regular character of G, and the multiplicity can be computed as the difference of the multiplicities of the trivial character 1_G. □

REMARK 30.13. Note that

$$
\tilde{\psi}_{G,h} + \overline{\tilde{\psi}_{G,h}} = \sum_{i=1}^{|h|-1}\left(\frac{i}{|h|}+\frac{|h|-i}{|h|}\right)\left(\varepsilon_h^{-i}\right)^G = \left(\sum_{i=1}^{|h|-1}\varepsilon_h^{-i}\right)^G = \left(\rho_{\langle h\rangle}-1_{\langle h\rangle}\right)^G,
$$

so the $\tilde{\psi}_{G,h}$ are candidates for "splitting the right hand side of the formula in Corollary 22.5 into a sum of two complex conjugate class functions". Theorem 30.12 shows that they are compatible with the Eichler Trace Formula

(Theorem 12.1). We could have proceeded the other way round, by obtaining the statement of Theorem 30.12 in a proof of the Eichler Trace Formula similar to that given in [**FK92**, V.2.4–V.2.9], and then deriving Theorem 12.1 from this equality.

REMARK 30.14. Not every $(\widetilde{\mathrm{RH}})$-class function of G is actually a character. Additionally to the possibility of constituents with negative multiplicity (see Remark 22.7), the multiplicities can be nonintegral, due to the denominators in $\widetilde{\psi}_{G,h}$.

EXAMPLE 30.15 (cf. Example 22.9). We show that the restriction of an $(\widetilde{\mathrm{RH}})$-character χ of G to a subgroup $U \leq G$ is an $(\widetilde{\mathrm{RH}})$-character of U.

Let χ be a G-character that satisfies the condition $(\widetilde{\mathrm{E}})$, and

$$\chi - 1_G = (g_0 - 1) \cdot \rho_G + \sum_{h \in G^\times / \sim_G} c_h \cdot \widetilde{\psi}_{G,h}.$$

Restriction to U yields

$$\chi_U - 1_U = (g_0 - 1)[G{:}U] \cdot \rho_G + \sum_{h \in G^\times / \sim_G} c_h \cdot \left(\widetilde{\psi}_{G,h} \right)_U.$$

By Mackey's Theorem (see page 24),

$$\left(\widetilde{\psi}_{G,h} \right)_U = \sum_{i=1}^{|h|-1} \frac{i}{|h|} \left(\left(\varepsilon_h^{-i} \right)^G \right)_U$$

$$= \sum_{i=1}^{|h|-1} \frac{i}{|h|} \sum_{t \in T} \left(\left(\varepsilon_{h^t}^{-i} \right)_{\langle h \rangle^t \cap U} \right)^U = \sum_{t \in T} \sum_{i=1}^{|h|-1} \frac{i}{|h|} \left(\left((\varepsilon_{h^t})_{\langle h \rangle^t \cap U} \right)^{-i} \right)^U,$$

where T is a set of representatives of $(\langle h \rangle, U)$–double cosets in G. The inner sum is of the form $\widetilde{\psi}_{U,k}$ for an appropriate k with $\langle k \rangle = \langle h \rangle \cap U$, so $\left(\widetilde{\psi}_{G,h} \right)_U$ is a sum of certain $\widetilde{\psi}_{U,k}$. Now $\varepsilon_h^U = \varepsilon_\sigma^U$ for $h \sim_U \sigma$ implies $\widetilde{\psi}_{U,h} = \widetilde{\psi}_{U,\sigma}$, thus we may choose k from a fixed set $U/\sim_U$, and reorder the summation to get

$$\chi_U - 1_U = (g_0' - 1) \cdot \rho_U + \sum_{k \in U^\times / \sim_U} c_k' \cdot \widetilde{\psi}_{U,k},$$

where the coefficients c_k' are sums of multiples of the c_h, and $g_0' - 1$ is a sum of $(g_0 - 1) \cdot [G{:}U]$ and multiples of the c_h. Now the claim follows. ◇

30.4. Compatibility of Characters and Homomorphisms.

As in Section 22.3, we ask again the natural question what the necessary conditions on the character χ mean for the existence of surface kernel epimorphisms Φ with $\mathrm{Tr}(\Phi) = \chi$. The character table of G suffices to decide the existence of a homomorphism $\Phi \colon \Gamma(\chi) \to G$ which maps the elliptic generators of

$\Gamma(\chi)$ to the conjugacy classes of G compatibly with an admissible system $(l_{\chi,u}(h))_{h \in G^\times, u \in I(|h|)}$. The surjectivity question is dealt with in the following lemma.

LEMMA 30.16 (cf. Lemma 22.11). *Let* $(l_{\chi,u}(h))_{h \in G^\times, u \in I(|h|)}$ *be an admissible system for the* $\widetilde{(RH)}$*-character* χ *of* G, C *a class structure of* G *in which the conjugacy class* h^G *occurs exactly* $l_{\chi,1}(h)$ *times, and* $\Phi \in \mathrm{Hom}_C(\Gamma(\chi), G)$ *with* $\Phi(\Gamma(\chi)) = U \le G$. *Then*

$$(\mathrm{Tr}(\Phi) - 1_U)^G = \chi - 1_G.$$

Proof. Setting $l_{\mathrm{Tr}(\Phi),1}(\sigma)$ to be the number of elliptic generators of $\Gamma(\chi)$ that are mapped to a U-conjugate of σ under Φ clearly defines an admissible system $(l_{\mathrm{Tr}(\Phi),u}(\sigma))_{\sigma \in U^\times, u \in I(|\sigma|)}$ for $\mathrm{Tr}(\Phi)$. The fusion of U-classes in G yields

$$l_{\chi,1}(h) = \sum_{\substack{\sigma \in U^\times / \sim_U \\ h \sim_G \sigma}} l_{\mathrm{Tr}(\Phi),1}(\sigma)$$

for all $h \in G^\times$. With $g_0 = [\chi, 1_G] = [\mathrm{Tr}(\Phi), 1_U]$, we have

$$\chi - 1_G - (g_0 - 1) \cdot \rho_G = \sum_{h \in U^\times / \sim_G} l_{\chi,1}(h) \cdot \widetilde{\psi}_{G,h} = \sum_{h \in U^\times / \sim_G} l_{\chi,1}(h) \cdot \widetilde{\psi}_{U,h}^G$$

by Theorem 30.12. As mentioned in Example 30.15, $\varepsilon_h^G = \varepsilon_\sigma^G$ for $h \sim_G \sigma$ implies $\widetilde{\psi}_{U,h}^G = \widetilde{\psi}_{U,\sigma}^G$, thus we get

$$\sum_{h \in U^\times / \sim_G} l_{\chi,1}(h) \cdot \widetilde{\psi}_{U,h}^G = \sum_{h \in U^\times / \sim_G} \left(\sum_{\substack{\sigma \in U^\times / \sim_U \\ h \sim_G \sigma}} l_{\mathrm{Tr}(\Phi),1}(\sigma) \right) \cdot \widetilde{\psi}_{U,\sigma}^G$$

$$= \left(\sum_{\sigma \in U^\times / \sim_U} l_{\mathrm{Tr}(\Phi),1}(\sigma) \cdot \widetilde{\psi}_{U,\sigma} \right)^G,$$

which is equal to $\mathrm{Tr}(\Phi) - 1_U - (g_0 - 1) \cdot \rho_U$, again by Theorem 30.12, and we are done. $\square$

30.5. Linearity of the Condition $\widetilde{(RH)}$. Similarly to Corollary 22.5, Theorem 30.12 can be formulated as a linear relation between the values $l_{\chi,1}(h)$ of a G-class function that satisfies the condition $\widetilde{(E)}$, and the coefficients of the irreducible G-characters in the decomposition of χ.

LEMMA 30.17 (cf. Lemma 22.16). *Let* $l_{\chi,1}(1) = [\chi, 1_G]$ *for each class function* χ *of the group* G, *and define the vector* $\widetilde{v}$ *and the matrix* $\widetilde{T}$ *by*

$$\widetilde{v} = (\widetilde{v}_\varphi)_{\varphi \in \mathrm{Irr}(G)}, \quad \text{with} \quad \widetilde{v}_{1_G} = 0 \quad \text{and} \quad \widetilde{v}_\varphi = -\varphi(1) \quad \text{otherwise,}$$

and

$$\widetilde{T} = \left(\widetilde{T}_{h,\varphi}\right)_{h \in G/\sim_G, \varphi \in \mathrm{Irr}(G)}, \quad \text{with} \quad \widetilde{T}_{h,\varphi} = [\widetilde{\psi}_{G,h}, \varphi].$$

If $\chi = \sum_{\varphi \in \mathrm{Irr}(G)} a_\varphi \cdot \varphi$ *is a* G-*class function satisfying the condition* $(\widetilde{\mathrm{E}})$ *and* $(l_{\chi,u}(h))_{h \in G^\times, u \in I(|h|)}$ *is an admissible system for* χ *then*

$$\left(a_\varphi\right)_{\varphi \in \mathrm{Irr}(G)} = \widetilde{v} + \left(l_{\chi,1}(h)\right)_{h \in G/\sim_G} \cdot \widetilde{T}.$$

Proof. With $g_0 = [\chi, 1_G]$, we have

$$
\begin{aligned}
a_\varphi &= [\chi, \varphi] \\
&= [1_G + (g_0 - 1) \cdot \rho_G, \varphi] + \sum_{h \in G^\times/\sim_G} l_{\chi,1}(h) \cdot [\widetilde{\psi}_{G,h}, \varphi] \\
&= [1_G - \rho_G, \varphi] + \sum_{h \in G/\sim_G} l_{\chi,1}(h) \cdot [\widetilde{\psi}_{G,h}, \varphi] \\
&= \widetilde{v}_\varphi + \sum_{h \in G/\sim_G} l_{\chi,1}(h) \cdot [\widetilde{\psi}_{G,h}, \varphi].
\end{aligned}
$$

$\square$

EXAMPLE 30.18 (cf. Example 22.18). Let $\varphi \in \mathrm{Irr}(G)$ and $h \in G$. Clearly $\widetilde{T}_{1,\varphi} = \varphi(1) \neq 0$ if $h = 1$, and for $h \in G^\times$ we get

$$\widetilde{T}_{h,\varphi} = [\widetilde{\psi}_{G,h}, \varphi] = \sum_{i=1}^{|h|-1} \frac{i}{|h|} [(\varepsilon_h^{-i})^G, \varphi] = \sum_{i=1}^{|h|-1} \frac{i}{|h|} [\varepsilon_h^{-i}, \varphi_{\langle h \rangle}],$$

which is zero if and only if $\langle h \rangle \subseteq \ker(\varphi)$. $\diamond$

EXAMPLE 30.19. As in Example 22.20, we consider the columns of $\widetilde{T}$ that belong to *linear* characters φ of G. We have $\widetilde{T}_{1,\varphi} = [\rho_G, \varphi] = 1$, and for $h \in G^\times$ with $\varphi(h) = \zeta_{|h|}^{-k}$, $0 \leq k < |h|$, we compute

$$\widetilde{T}_{h,\varphi} = [\widetilde{\psi}_{G,h}, \varphi] = \sum_{i=1}^{|h|-1} \frac{i}{|h|} [(\varepsilon_h^{-i})^G, \varphi] = \sum_{i=1}^{|h|-1} \frac{i}{|h|} [\varepsilon_h^{-i}, \varphi_{\langle h \rangle}] = \frac{k}{|h|}.$$

    $\diamond$

EXAMPLE 30.20. As stated already after Definition 30.8, there are in general different admissible systems for a class function that satisfies the condition $(\widetilde{\mathrm{E}})$. For $G = \langle a \mid a^5 \rangle$, the cyclic group of order 5, consider the character $\chi = 1_G + \widetilde{\psi}_{G,a} + \widetilde{\psi}_{G,a^{-1}}$. By Remark 30.13 and $\overline{\widetilde{\psi}_{G,h}} = \widetilde{\psi}_{G,h^{-1}}$, we have $\chi = \rho_G = 1_G + \widetilde{\psi}_{G,a^2} + \widetilde{\psi}_{G,a^{-2}}$. $\diamond$

EXAMPLE 30.21. Let $G = \langle h \rangle$ be a cyclic group of order $m > 1$. We define $\varphi_j \in \mathrm{Irr}(G)$, $0 \le j < m$, by $\varphi_j(h) = \zeta_m^{-j}$, and denote, for $u \in \mathbb{Z}$, by $R(u)$ the unique integer with $R(u) \equiv u \pmod{m}$ and $0 \le R(u) < m$. Then $\widetilde{T}_{h^u,\varphi_j} = k/|h^u|$, where $\varphi_j(h^u) = \zeta_{|h^u|}^{-k}$. With $d = m/|h^u|$, we have $\varphi_j(h^u) = \zeta_m^{-uj} = \zeta_{|h^u|}^{-uj/d}$ and thus $k \equiv uj/d \pmod{m/d}$. This means $\widetilde{T}_{h^u,\varphi_j} = k/|h^u| = R(uj)/m$.

So the matrices $m \cdot \widetilde{T}$ for $m \in \{2,3,4,5\}$ are the following.

$$
\begin{pmatrix} 2 & 2 \\ 0 & 1 \end{pmatrix}
\quad
\begin{pmatrix} 3 & 3 & 3 \\ 0 & 1 & 2 \\ 0 & 2 & 1 \end{pmatrix}
\quad
\begin{pmatrix} 4 & 4 & 4 & 4 \\ 0 & 1 & 2 & 3 \\ 0 & 2 & 0 & 2 \\ 0 & 3 & 2 & 1 \end{pmatrix}
\quad
\begin{pmatrix} 5 & 5 & 5 & 5 & 5 \\ 0 & 1 & 2 & 3 & 4 \\ 0 & 2 & 4 & 1 & 3 \\ 0 & 3 & 1 & 4 & 2 \\ 0 & 4 & 3 & 2 & 1 \end{pmatrix}
$$

The matrices for $m < 5$ are invertible, the rank of the matrix for $m = 5$ is 4, due to the general property that the sum of the rows for h and h^{-1} is equal to the sum of the rows for h^2 and h^{-2}, see Example 30.20.   ⋄

30.6. Ambiguities in the Choice of $l_{\chi,u}(h)$. Unlike the matrix T introduced in Lemma 22.16, the matrix $\widetilde{T}$ from Lemma 30.17 is in general not invertible. This corresponds to the ambiguity in the choice of admissible systems for a given class function, see Example 30.20. More specifically, we have the following.

LEMMA 30.22. *Let T and $\widetilde{T}$ be defined as in Lemma 22.16 and Lemma 30.17, respectively, for the group G. Let $\varphi \in \mathrm{Irr}(G)$ and $\vartheta \in \mathrm{Rat}(G)$ with $[\varphi, \vartheta] \neq 0$. Then the following statements hold.*

(a) $\widetilde{T}_{h,\varphi} = \widetilde{T}_{h^{-1},\overline{\varphi}}$.

(b) $\widetilde{T}_{h,\varphi} + \widetilde{T}_{h,\overline{\varphi}} = T_{\langle h \rangle, \vartheta}$.

(c) *If $\varphi \in \mathrm{Irr}(G)$ with $\varphi = \overline{\varphi}$ then $\widetilde{T}_{h,\varphi} = \widetilde{T}_{h,\varphi^{*k}}$ for all $h \in G$ and all Galois conjugates φ^{*k}.*

(d) *If $h \in G$ with $h \sim_G h^{-1}$ then $\widetilde{T}_{h^k,\varphi} = \widetilde{T}_{h,\varphi}$ for all $\varphi \in \mathrm{Irr}(G)$ and all integers k with $\gcd(k, |h|) = 1$.*

Proof. Part (a) follows from $\varepsilon_{h^{-1}} = \overline{\varepsilon_h}$ and thus $\widetilde{\psi}_{G,h^{-1}} = \overline{\widetilde{\psi}_{G,h}}$.

For part (b), consider $\widetilde{\psi}_{G,h} + \overline{\widetilde{\psi}_{G,h}} = \psi_{G,\langle h \rangle}$ (see Remark 30.13), so

$$\widetilde{T}_{h,\varphi} + \widetilde{T}_{h,\overline{\varphi}} = \left[\widetilde{\psi}_{G,h} + \overline{\widetilde{\psi}_{G,h}}, \varphi \right] = \left[\psi_{G,\langle h \rangle}, \varphi \right] = \left[\psi_{G,\langle h \rangle}, \vartheta \right] / [\vartheta, \vartheta] = T_{\langle h \rangle, \vartheta}.$$

In part (c), we may assume without loss of generality that k is coprime to $|G|$; if $\overline{k}$ denotes an integer with $k\overline{k} \equiv 1 \pmod{|G|}$ then $[\psi, \chi^{*k}] = [\psi^{*\overline{k}}, \chi]$

for any two class functions ψ, χ of G. In particular, if $\varphi = \overline{\varphi}$ then $\varphi^{*k} = \overline{\varphi^{*k}}$, and thus $[\widetilde{\psi}_{G,h}, \varphi^{*k}] = [\widetilde{\psi}_{G,h}, \overline{\varphi^{*k}}] = [\overline{\widetilde{\psi}_{G,h}}, \varphi^{*k}]$ implies

$$2 \cdot [\widetilde{\psi}_{G,h}, \varphi^{*k}] =$$

$$[\widetilde{\psi}_{G,h} + \overline{\widetilde{\psi}_{G,h}}, \varphi^{*k}] = [(\rho_{\langle h\rangle} - 1_{\langle h\rangle}), \varphi^{*k}] = [(\rho_{\langle h\rangle} - 1_{\langle h\rangle}), \varphi] = 2 \cdot [\widetilde{\psi}_{G,h}, \varphi].$$

To show part (d), note that $h^k \sim_G h^{-k}$, so $\widetilde{T}_{h^k,\varphi} = \widetilde{T}_{h^{-k},\varphi} = \widetilde{T}_{h^k,\overline{\varphi}}$ holds by part (a), and thus

$$2 \cdot \widetilde{T}_{h^k,\varphi} = \widetilde{T}_{h^k,\varphi} + \widetilde{T}_{h^k,\overline{\varphi}} = T_{\langle h^k\rangle,\vartheta} = T_{\langle h\rangle,\vartheta} = 2 \cdot \widetilde{T}_{h,\varphi}$$

by part (b) and $\langle h^k\rangle = \langle h\rangle$; we see that the rows of h and h^k in $\widetilde{T}$ are equal. $\qquad\square$

So we may identify in $\widetilde{T}$ the columns of characters φ and φ^{*k} for $\varphi = \overline{\varphi}$ and the rows of h and h^k for $h \sim_G h^{-1}$ and $\gcd(k, |h|) = 1$. The collapsed matrix $\hat{T}$ is then defined as follows.

DEFINITION 30.23. For a finite group G, let $S(G)$ denote a set of representatives of certain G-conjugacy classes, obtained by taking exactly one representative h for each *nonreal* class h^G, and taking exactly one representative h for the rational class of h if h is *real*. Let $P(G)$ denote the union of the set of all nonreal characters in $\mathrm{Irr}(G)$ with the set of all those characters in $\mathrm{Rat}(G)$ whose constituents are all real. Finally, set

$$\hat{T} = \left(\widetilde{T}_{h,\varphi}\right)_{h\in S(G), \varphi\in P(G)}.$$

EXAMPLE 30.24.   (a) If $|G|$ is odd then the only real irreducible character of G is 1_G, thus $S(G) = G/\sim_G$, $P(G) = \mathrm{Irr}(G)$, and $\hat{T} = \widetilde{T}$.

(b) G is ambivalent if and only if we can take $S(G) = \widehat{CY}(G)/\sim_G$ and $P(G) = \mathrm{Rat}(G)$, and this is equivalent to $2 \cdot \hat{T} = T$.

(c) All characters of G are rational if and only if $\mathrm{Irr}(G) = \mathrm{Rat}(G)$, which is equivalent to $2 \cdot \hat{T} = 2 \cdot \widetilde{T} = T$. $\diamond$

Under the assumptions of Lemma 30.10, we can characterize the situation that $\hat{T}$ is invertible. First we note that the system of equations

$$(a_\varphi)_{\varphi\in\mathrm{Irr}(G)} = \widetilde{v} + (l_{\chi,1}(h))_{h\in G/\sim_G} \cdot \widetilde{T}$$

from Lemma 30.17 can be formulated as

$$(a_\varphi)_{\varphi\in P(G)} = (\widetilde{v}_\varphi)_{\varphi\in P(G)} + \left(\hat{l}_\chi(h)\right)_{h\in S(G)} \cdot \hat{T}.$$

Here the collapsed vector $\hat{l}_\chi = (\hat{l}_\chi(h))_{h\in S(G)}$ is defined by $\hat{l}_\chi(h) = l_{\chi,1}(h)$ if $h \not\sim_G h^{-1}$, and $\hat{l}_\chi(h) = \sum_k l_{\chi,1}(h^k)$ otherwise, where the sum is taken over the different G-classes of generators h^k of $\langle h\rangle$.

So $\hat{T}$ is invertible if and only if $\hat{l}_\chi$ is uniquely determined by χ.

LEMMA 30.25. *Let G be a finite group such that for all $h \in G$ with $|h| > 2$, $|h|$ has the property* (LI). *Then $\hat{T}$ is invertible if and only if for each $h \in G$ with $h \not\sim_G h^{-1}$, the generators of $\langle h \rangle$ lie in two G-conjugacy classes.*

Proof. First suppose that h, h^{-1}, h^k, and h^{-k} are pairwise nonconjugate in G, with $\langle h \rangle = \langle h^k \rangle$; then four different rows of $\hat{T}$ correspond to these elements. Since $\widetilde{T}_{h,\varphi} + \widetilde{T}_{h^{-1},\varphi} = \widetilde{T}_{h^k,\varphi} + \widetilde{T}_{h^{-k},\varphi}$ for all $\varphi \in P(G)$, by Lemma 30.22, the matrix $\hat{T}$ is not invertible.

So let us now assume that for each $h \in G$, either $h \sim_G h^{-1}$ holds or the generators of $\langle h \rangle$ lie in two G-conjugacy classes; the latter situation occurs if and only if $\mathbb{Q}(\{\chi(h) \mid \chi \in \mathrm{Irr}(G)\})$ is a quadratic extension of $\mathbb{Q}$. Let $(l_{\chi,u}(h))_{h \in G^\times, u \in I(|h|)}$ be an admissible system for χ.

If $h \sim_G h^{-1}$ then clearly $\hat{l}_\chi(h) = \sum_k l_{\chi,1}(h^k) = l_\chi(\langle h \rangle)$ (see the formula before Lemma 30.10) is determined by χ. If $h \not\sim_G h^{-1}$ then $\hat{l}_\chi(h) = l_{\chi,1}(h)$; by Lemma 30.10, $l_{\chi,1}(h) - l_{\chi,1}(h^{-1})$ can be computed from χ, and together with $l_\chi(\langle h \rangle) = l_{\chi,1}(h) + l_{\chi,1}(h^{-1})$, we can compute $l_{\chi,1}(h)$. $\square$

31. Finiteness Results

Now we discuss the third question from the introduction to Section 23, namely to decide for a given G-character χ whether there is a surface kernel epimorphism Φ with $\mathrm{Tr}(\Phi) = \chi$.

31.1. Classification for Large $[\chi, 1_G]$.

In Section 23.1, the crucial step is to show that for $\Phi \in \mathrm{Hom}_C(\Gamma(\chi), G)$, with G ambivalent, the product $\sigma = \prod_{i=1}^r \Phi(c_i)$ of images of the elliptic generators lies in G'. Since G/G' is an ambivalent abelian group in this situation, it must in fact be an elementary abelian 2-group, and it is sufficient to prove that σ lies in each normal subgroup of index 2 in G. In the general case, the argumentation is analogous. G/G' is an abelian group, and we show that σ lies in each normal subgroup of G with cyclic factor group. Note that G' is the intersection of these normal subgroups.

LEMMA 31.1 (cf. Lemma 23.1). *Let G be a group, N a normal subgroup of G with G/N cyclic, χ an $(\widetilde{\mathrm{RH}})$-character of G, $(l_{\chi,u}(h))_{h \in G^\times, u \in I(|h|)}$ an admissible system for χ, $C = (C_1, C_2, \ldots, C_r)$ a class structure of G in which the class h^G occurs exactly $l_{\chi,1}(h)$ times, and $\sigma_i \in C_i$ for $1 \le i \le r$. Then $\prod_{i=1}^r \sigma_i \in N$.*

Proof. Let $\pi\colon G \to \langle h \rangle$ be an epimorphism onto a cyclic group of order $m > 1$, with $N = \ker(\pi)$. Consider the linear character $\varphi = \varepsilon_h^{-1} \circ \pi \in \mathrm{Irr}(G)$. With the notation of Lemma 30.17, we get

$$a_\varphi - \widetilde{v}_\varphi = \sum_{\sigma \in G/\sim_G} l_{\chi,1}(\sigma) \cdot \widetilde{T}_{\sigma,\varphi} = \sum_{d\mid m} \sum_{j \in I(m/d)} \sum_{\substack{\sigma \in G/\sim_G \\ \pi(\sigma)=h^{dj}}} l_{\chi,1}(\sigma) \cdot \widetilde{T}_{\sigma,\varphi}.$$

(Here we set $I(1) = \{1\}$; but the summand for $d = m$ is unimportant for the argument below.)

If $\pi(\sigma) = h^{dj}$ then $\varphi(\sigma) = \varepsilon_h^{-1}(\pi(\sigma)) = \zeta_m^{-dj}$, so $\widetilde{T}_{\sigma,\varphi} = dj/m$, independently of $|\sigma|$. This yields

$$\begin{aligned}
a_\varphi - \widetilde{v}_\varphi &= \frac{1}{m} \sum_{d\mid m} \sum_{j \in I(m/d)} dj \sum_{\substack{\sigma \in G/\sim_G \\ \pi(\sigma)=h^{dj}}} l_{\chi,1}(\sigma) \\
&= \frac{1}{m} \sum_{d\mid m} \sum_{j \in I(m/d)} dj \cdot |\{ i \mid 1 \le i \le r, \pi(\sigma_i) = h^{dj} \}|.
\end{aligned}$$

Now we see that

$$\pi\left(\prod_{i=1}^r \sigma_i \right) = \prod_{d\mid m} \prod_{j \in I(m/d)} \prod_{\substack{1 \le i \le r \\ \pi(\sigma_i)=h^{dj}}} \pi(\sigma_i) = \prod_{d\mid m} \prod_{j \in I(m/d)} \prod_{\substack{1 \le i \le r \\ \pi(\sigma_i)=h^{dj}}} h^{dj}$$

is a power of h, the exponent being equal to $m \cdot (a_\varphi - \widetilde{v}_\varphi)$ by the above calculation. So the product is indeed the identity of $\langle h \rangle$. $\square$

With this lemma, the following generalization of Theorem 23.2 is obvious.

THEOREM 31.2. *Let G be a finite group that can be generated by n elements. Let χ be an $\widetilde{(\mathrm{RH})}$-character of G such that $[\chi, 1_G] \ge c.l.(G) + n/2$. Then there is a $\Phi \in \mathrm{Epi}_{L(\chi)}(\Gamma(\chi), G)$ such that $\mathrm{Tr}(\Phi) = \chi$.*

Proof. The idea is the same as in the proof of Theorem 23.2. Note that we may choose the elements $\sigma_1, \sigma_2, \ldots, \sigma_r$ needed in this proof in such a way that $l_{\chi,1}(h)$ of them are G-conjugate to h, for all $h \in G^\times$, where $(l_{\chi,u}(h))_{h \in G^\times, u \in I(|h|)}$ is an admissible system for χ. This yields $\mathrm{Tr}(\Phi) = \chi$. $\square$

31.2. Infinite Numbers of Exceptions. We revisit the examples of Section 23.2.

EXAMPLE 31.3. In the situation of Theorem 23.5, let $Z = \langle z \rangle$. Then $\chi = 1_G + n \cdot \widetilde{\psi}_{G,z} + n \cdot \widetilde{\psi}_{G,z^{-1}}$ by Remark 30.13, so χ is an $\widetilde{(\mathrm{RH})}$-character. $\diamond$

EXAMPLE 31.4. In the situation of Theorem 23.8, the character χ also clearly does not come from a Riemann surface if we assume additionally that χ is an $\widetilde{(\text{RH})}$-character.   ◇

We see that the condition $\widetilde{(\text{RH})}$ does not exclude the two examples of infinite series of (RH)-characters that do not come from Riemann surfaces. Now let us look at a third example.

THEOREM 31.5. *Let G be a group that cannot be generated by two elements, such that G contains an element σ of order p^d, for a prime p and $d > 1$. Furthermore, assume that $U = \langle \sigma^p \rangle$ is normal in G. Then any character of the form*

$$\chi = 1_G + \sum_{h \in U^\times / \sim_G} l_{\chi,1}(h) \cdot \widetilde{\psi}_{G,h},$$

with nonnegative integers $l_{\chi,1}(h)$ that are not all zero, satisfies the condition $\widetilde{(\text{RH})}$ but does not come from a Riemann surface.

Proof. Suppose $\Phi \colon \Gamma(\chi) \to G$ is a surface kernel epimorphism, and let h_1, h_2 be the images of the hyperbolic generators. Then $\Phi(\Gamma(\chi)) \subseteq \langle h_1, h_2, U \rangle$. Consider the natural epimorphism from G to G/U. Its image is generated by $h_1 U$ and $h_2 U$, and since σU lies in the image, there is an element $\tau \in \langle h_1, h_2 \rangle$ that is mapped to σU. In other words, $\tau = \sigma^{1+pk}$ for some integer k, and thus $\langle \tau \rangle = \langle \sigma \rangle \leq \langle h_1, h_2 \rangle$. So $G = \Phi(\Gamma(\chi)) = \langle h_1, h_2 \rangle$, contrary to the assumption that G cannot be generated by two elements.   □

REMARK 31.6. In [**KK90b**, Theorem on p. 290], it is shown that a character of degree 2, 3, or 4 comes from a Riemann surface if and only if it satisfies the conditions $\widetilde{(\text{E})}$ and $\widetilde{(\text{RH})}$. This follows from the inspection of all automorphism groups in these genera. For genus 5, that is, for character degree 5, the conditions $\widetilde{(\text{E})}$ and $\widetilde{(\text{RH})}$ are not sufficient. The character $1_G + 2\chi_5$ of the dihedral group D_8 is a counterexample (see Example 23.6).

31.3. Finite Numbers of Exceptions. We try to generalize the results of Section 23.3. For this, we first generalize the necessary terminology in a straightforward manner.

DEFINITION 31.7. For a group G and $\sigma \in G^\times$, let $\widetilde{rep}(\sigma^G)$ denote the smallest integer such that for all $m \geq \widetilde{rep}(\sigma^G)$, σ can be written as a product of exactly m elements in the conjugacy class σ^G; if no such number exists then we set $\widetilde{rep}(\sigma^G) = \infty$.

We define $\widetilde{n}(\sigma)$ to be the smallest integer n such that each element in $\langle G', \sigma \rangle$ can be written as a product of at most n elements in σ^G. If no such integer exists then we set $\widetilde{n}(\sigma) = \infty$.

We define $\widetilde{k}(\sigma)$ to be the smallest integer such that $\langle G', \sigma \rangle$ can be generated by $\widetilde{k}(\sigma)$ elements in σ^G. If no such integer exists then we set $\widetilde{k}(\sigma) = \infty$.

If G is a group such that G' is simple and contained in every nontrivial proper normal subgroup of G then the conjugacy class σ^G generates a group containing $\langle G', \sigma \rangle$ for any $\sigma \in G^\times$, hence $\widetilde{n}(\sigma)$ and $\widetilde{k}(\sigma)$ are finite in this case.

$\widetilde{rep}(\sigma^G)$ and $\widetilde{n}(\sigma)$ can be computed from the character table of G. In general this does not hold for $\widetilde{k}(\sigma)$.

THEOREM 31.8 (cf. Theorem 23.13). *Let G be a finite simple group, and assume that $\widetilde{rep}(\sigma^G) < \infty$ for all $\sigma \in G^\times$. Then at most finitely many $(\widetilde{RH})$-characters of G do not come from Riemann surfaces.*

Proof. Let χ be an $(\widetilde{RH})$-character of G and $(l_{\chi,u}(h))_{h \in G^\times, u \in I(|h|)}$ an admissible system for χ such that $l_{\chi,1}(\sigma) \geq \widetilde{n}(\sigma) + \widetilde{k}(\sigma) + \widetilde{rep}(\sigma^G) - 1$ holds for an element $\sigma \in G$. We claim that there is a $\Phi \in \mathrm{Epi}_{L(\chi)}(\Gamma(\chi), G)$ such that $\chi = \mathrm{Tr}(\Phi)$.

To define Φ, we choose the identity of G as image of the hyperbolic generators of $\Gamma(\chi)$, and choose arbitrary elements $\tau_{i,k} \in k^G$, $1 \leq i \leq l_{\chi,1}(k)$, for $k \in G^\times / \sim_G$ with $k \not\sim_G \sigma$. Then we set

$$\tau = \prod_{\substack{k \in G^\times / \sim_G \\ k \not\sim_G \sigma}} \prod_{i=1}^{l_{\chi,1}(k)} \tau_{i,k},$$

and choose $k = \widetilde{k}(\sigma)$ elements $\sigma_1, \sigma_2, \ldots, \sigma_k$ in σ^G that generate G. Then $(\tau \sigma_1 \sigma_2 \cdots \sigma_k)^{-1}$ is a product of $m \leq \widetilde{n}(\sigma)$ elements $\sigma_{k+1}, \sigma_{k+2}, \ldots, \sigma_{k+m}$ in σ^G, so $\tau \sigma_1 \sigma_2 \cdots \sigma_{k+m}$ is the identity.

There are still $l = l_{\chi,1}(\sigma) - k - m$ elements in σ^G to be defined, and because $l \geq \widetilde{rep}(\sigma^G) - 1$, we can replace σ_{k+m} by $l + 1$ elements $\sigma'_1, \sigma'_2, \ldots, \sigma'_{l+1}$ in its conjugacy class. Choosing the $\tau_{i,k}$, σ_i $(1 \leq i \leq k + m - 1)$, and σ'_i $(1 \leq i \leq l + 1)$ as images of the elliptic generators of $\Gamma(\chi)$ under Φ, we get $\Phi \in \mathrm{Epi}_{L(\chi)}(\Gamma(\chi), G)$ with $\mathrm{Tr}(\Phi) = \chi$.

Now the statement of the theorem follows from the facts that only finitely many values of $g_0 = [\chi, 1_G]$ must be considered by Theorem 31.2, and that for each of them, at most finitely many vectors $(l_{\chi,1}(k))_{k \in G^\times / \sim_G}$ do not admit the construction of Φ as above. $\square$

Now we generalize the idea of Theorem 31.8 analogously to Theorem 23.13. We have $\widetilde{rep}(\sigma^G) = \infty$ for all $\sigma \in G \setminus G'$ because $\sigma = \sigma_1 \sigma_2 \cdots \sigma_n$ for $\sigma_i \in \sigma^G$ implies $\sigma G' = (\sigma_1 G')(\sigma_2 G') \cdots (\sigma_n G') = (\sigma G')^n$, hence $n - 1$ is divisible by the order of $\sigma G'$ in the group G/G'. As in Section 23.3, the finiteness condition on $\widetilde{rep}(\sigma^G)$ is needed only for $\sigma \in G'$.

THEOREM 31.9 (cf. Theorem 23.15). *Let G be a finite group such that G' is a simple group that is contained in every nontrivial proper normal subgroup of G. Assume that $\widetilde{rep}(\sigma^{G'}) < \infty$ for all $\sigma \in G'$.*

Further assume that each $\sigma \in G \setminus G'$ satisfies $\mathrm{Hom}_{C(\sigma)}(0, G) \neq \emptyset$, where $C(\sigma)$ is the class structure that consists of $|\sigma G'|$ times the conjugacy class of σ.

Then all except at most finitely many (RH)-characters χ of G either with the property that $g_0 = [\chi, 1_G]$ is zero or such that G/G' can be generated by $2g_0$ elements come from Riemann surfaces.

In particular, if G can be generated by two elements then at most finitely many (RH)-characters of G do not come from Riemann surfaces.

Proof. First let χ be an $\widetilde{(RH)}$-character of G and $(l_{\chi,u}(h))_{h \in G^\times, u \in I(|h|)}$ an admissible system for χ such that $l_{\chi,1}(\sigma) \geq \tilde{n}(\sigma) + \tilde{k}(\sigma) + \widetilde{rep}(\sigma^G) - 1$ holds for an element $\sigma \in G^\times$. Then we can construct the homomorphism Φ as in the proof of Theorem 31.8. Note that the product of the images of the elliptic generators lies in G' by Lemma 31.1. The surjectivity of Φ for $g_0 = 0$ follows from Lemma 22.19. If G/G' can be generated by $2g_0$ elements then we choose images of the $2g_0$ hyperbolic generators that generate G modulo G'. Now let χ be an $\widetilde{(RH)}$-character of G and and $(l_{\chi,u}(h))_{h \in G^\times, u \in I(|h|)}$ an admissible system for χ such that $l_{\chi,1}(\sigma) \geq \tilde{n}(\sigma) + \tilde{k}(\sigma) + \widetilde{rep}(\sigma^G) - 1$ holds for an element $\sigma \in G \setminus G'$. Similarly to the proofs of Theorem 31.8 and of the case above, we define a homomorphism $\Phi \colon \Gamma(\chi) \to G$. We choose arbitrary images in the appropriate conjugacy classes for those elliptic generators of $\Gamma(\chi)$ that are mapped to elements $k \not\sim_G \sigma$; let τ denote the product of these elements. By Lemma 31.1, we have $\tau \in \langle G', \sigma \rangle$. Next we choose $k = \tilde{k}(\sigma)$ images $\sigma_1, \sigma_2, \ldots, \sigma_k$ in σ^G that generate $\langle G', \sigma \rangle$. Then $(\tau \sigma_1 \sigma_2 \cdots \sigma_k)^{-1}$ lies in $\langle G', \sigma \rangle$, so is a product of $m \leq \tilde{n}(\sigma)$ elements $\sigma_{k+1}, \sigma_{k+2}, \ldots, \sigma_{k+m}$ in σ^G. Hence $\tau \sigma_1 \sigma_2 \cdots \sigma_{k+m}$ is the identity. Now $l = l_{\chi,1}(\sigma) - k - m$ images in σ^G are left to be defined. Because of Lemma 31.1, l is a multiple of the order of $\sigma G'$ in G/G', and by the assumption, we may choose l elements in σ^G such that their product is the identity. Thus we get a homomorphism Φ, which is surjective for $g_0 = 0$ by Lemma 22.19. If G/G' can be generated by $2g_0$ elements then surjectivity can be achieved by an appropriate choice of images of the hyperbolic generators of $\Gamma(\chi)$.

Now the statement of the theorem follows as in the proof of Theorem 31.8. □

REMARK 31.10. The additional condition that $\mathrm{Hom}_{C(\sigma)}(0, G) \neq \emptyset$ is trivially satisfied if the order of σ in G is equal to the order of $\sigma G'$ in G/G', in particular if σ has prime order.

Note that if G is ambivalent, the condition is satisfied because the order of $\sigma G'$ in G/G' is 2 for $\sigma \notin G'$, and $\sigma \sim_G \sigma^{-1}$ implies $\operatorname{Hom}_{(\sigma^G, \sigma^G)}(0, G) \neq \emptyset$.

32. Abelian Groups of Automorphisms

Now we generalize the example discussed in Section 25. Let G be a finite *abelian* group, χ an $(\widetilde{\mathrm{RH}})$-character of G, $(l_{\chi,u}(h))_{h \in G^\times, u \in I(|h|)}$ an admissible system for χ, and C a class structure in which the class h^G occurs exactly $l_{\chi,1}(h)$ times, Lemma 31.1 guarantees that $\operatorname{Hom}_C(\Gamma(\chi), G)$ is nonempty. So the question whether χ comes from a Riemann surface is just the question whether a *surjective* surface kernel homomorphism exists. We can formulate this as an analogue of Theorem 25.1, as follows.

THEOREM 32.1. *A character χ of a finite abelian group G comes from a Riemann surface if and only if it satisfies the condition $(\widetilde{\mathrm{RH}})$ and an admissible system $(l_{\chi,u}(h))_{h \in G^\times, u \in I(|h|)}$ exists such that G can be generated by $\{h \in G^\times \mid l_{\chi,1}(h) \neq 0\}$ together with $2g_0$ arbitrary elements of G.*

COROLLARY 32.2. *If G is abelian and can be generated by at most two elements then the condition $(\widetilde{\mathrm{RH}})$ is necessary and sufficient for a character of G to come from a Riemann surface.*

Proof. By Lemma 22.19, any surface kernel homomorphism corresponding to an $(\widetilde{\mathrm{RH}})$-character χ with $[\chi, 1_G] = 0$ is surjective. For $[\chi, 1_G] \geq 1$, surjectivity can be guaranteed by mapping the hyperbolic generators of $\Gamma(\chi)$ to generators of G. □

Note that by Theorem 31.5, the condition $(\widetilde{\mathrm{RH}})$ is in general *not* sufficient for characters of other abelian groups to come from Riemann surfaces.

REMARK 32.3. In [**Kur87**], the statement of Theorem 32.1 is proved for the special case of cyclic groups. In fact, the (very technical) proof of [**Kur87**, Lemma 2.13] uses an argument similar to that in the proof of Theorem 32.1. This result was used for example in [**KK91**] to introduce a necessary condition called (CY), which expresses that the restrictions of a G-character χ to all (maximal) cyclic subgroups of G come from Riemann surfaces. In our terminology, χ satisfies the condition (CY) if the restrictions to all cyclic subgroups of G are $(\widetilde{\mathrm{RH}})$-characters, which holds automatically if χ is an $(\widetilde{\mathrm{RH}})$-character, by Example 30.15.

REMARK 32.4. For a character χ of a cyclic group $G = \langle h \rangle$, the collection of numbers $r_{\chi,u}(h^d)$, d dividing $|G|$ and $u \in I(|h^d|)$, is called "rotation datum" in [**Kur87**], and if the underlying $l_{\chi,u}(h^d)$ are integral then the rotation datum is called "normal". So the existence theorem there, [**Kur87**, Proposition 4.5], states that a character χ of a cyclic group comes from a Riemann surface if and only if a normal rotation datum for χ exists.

COROLLARY 32.5. *Let $\langle h \rangle$ be a cyclic group of prime order p. A character χ of $\langle h \rangle$ comes from a Riemann surface if and only if $\chi(1) \geq 2$ and $\chi(h)$ satisfies the Eichler Trace Formula for p.*

Proof. We have $l_{\chi,1}(h) = r^*_{\chi,1}(h) = r_{\chi,1}(h)$, so the condition $(\widetilde{\mathrm{RH}})$ is not stronger than the condition $(\widetilde{\mathrm{E}})$, and this means exactly that $\chi(1) \geq 2$ and $\chi(h^u)$ satisfies the Eichler Trace Formula for p, for $1 \leq u \leq p - 1$. Since the values $\chi(h^u)$ are algebraic conjugates of $\chi(h)$, we have to check only $\chi(h)$. $\square$

REMARK 32.6. A direct proof of Corollary 32.5 is given in [**Kur83**]. There it is proved that $\alpha = \sum_{u=1}^{p-1} f_u \frac{\zeta_p^u}{1-\zeta_p^u}$, with nonnegative integers f_u, is an algebraic integer if and only if $\sum_{u=1}^{p-1} f_u \cdot u' \equiv 0 \pmod{p}$, where u' is defined by $u \cdot u' \equiv 1 \pmod{p}$.

It was shown already in [**Kur66**, 3.2 and 3.5] that this relation is sufficient for the existence of a Riemann surface X with an automorphism σ of prime order p, where the action of σ on X induces the character χ that is given by $\chi(1) = g(X)$ and $\chi(\sigma) = \alpha$. The proof given there uses explicit computations with rotation constants and function fields, not character theoretic methods.

33. A Classification Algorithm

For the sake of simplicity, we generalize the classification algorithm in Section 24 only to the situation of Lemma 30.25, that is, the matrix $\hat{T}$ of G is invertible; the examples in the following sections are all of this type. (Note that in the general case, an $(\widetilde{\mathrm{RH}})$-class function χ does not uniquely determine the vector $\hat{l}_\chi$. Thus a character χ is proved not to come from a Riemann surface if *all* corresponding vectors $\hat{l}_\chi$ are excluded; conversely, if one vector $\hat{l}_\chi$ establishes that χ comes from a Riemann surface, of course the other vectors for χ need not be tested.)

First we need an analogue of Lemma 24.1.

LEMMA 33.1. *Let $\hat{l} = (\hat{l}(h))_{h \in S(G)}$ be a vector of nonnegative integers. If the corresponding class function $\chi = \sum_{\varphi \in P(G)} a_\varphi \cdot \varphi$ with $(a_\varphi)_{\varphi \in P(G)} = (\widetilde{v}_\varphi)_{\varphi \in P(G)} + \hat{l} \cdot \hat{T}$ is not an $(\widetilde{\mathrm{RH}})$-character then at least one of the following cases occurs:*

(i) *a has a nonintegral entry;*
(ii) *a has a negative entry;*
(iii) *$\chi(1) < 2$.*

In Section 24, we frequently used the fact that increasing $l(H)$ by 2 maps each (RH)-vector l to another (RH)-vector. Now we define, for $h \in G^\times$, the number $m(h)$ as the smallest positive integer such that the identity can be written as the product of $m(h)$ elements in the class h^G. Clearly $m(h) = 2$ if $h \sim_G h^{-1}$, and always $m(h) \leq |h|$ holds. If χ is a G-character that comes from a Riemann surface then the underlying class function of the vector obtained from $\hat{l}_\chi$ by increasing $\hat{l}_\chi(h)$ by $m(h)$ also comes from a Riemann surface; in particular the denominator of $\widetilde{T}_{h,\varphi}$, for $\varphi \in \mathrm{Irr}(G)$, divides $m(h)$.

We say that l is an $\widetilde{(\mathrm{RH})}$-vector if the corresponding class function is an $\widetilde{(\mathrm{RH})}$-class function.

LEMMA 33.2. *Let G be a finite group, g_0 a nonnegative integer, and $h \in G^\times$ not contained in a proper normal subgroup of G. For a nonnegative integer q, we define $\hat{M}(q, h)$ as the set of those vectors $(\hat{l}(k))_{k \in S(G)}$ over the integers with the properties $\hat{l}(1) = g_0$, $q \leq \hat{l}(h) \leq q + m(h) - 1$, and $0 \leq l(k) \leq m(k) - 1$ for $k \not\sim_G h$.*

Suppose that each character corresponding to an $\widetilde{(\mathrm{RH})}$-vector in $\hat{M}(q, h)$ comes from a Riemann surface, and that no vector in $\hat{M}(q, h)$ fails to be an $\widetilde{(\mathrm{RH})}$-vector because of the condition 33.1 (iii). If $g_0 = 0$ then additionally assume that

$$q \geq \max\left\{ \frac{-\widetilde{v}_\varphi}{\widetilde{T}_{h,\varphi}} \middle| \varphi \in \mathrm{Irr}(G), \varphi \neq 1_G \right\}.$$

Then each character corresponding to an $\widetilde{(\mathrm{RH})}$-vector $\hat{l}$ with $\hat{l}(h) \geq q$ comes from a Riemann surface.

Proof. Let $\hat{l}$ be an $\widetilde{(\mathrm{RH})}$-vector with $\hat{l}(H) \geq q$, and consider the (unique) vector $\hat{l}' \in \hat{M}(q, h)$ that is given by the properties $\hat{l}'(k) \equiv \hat{l}(k) \pmod{m(k)}$ for all $k \in S(G)$.

We claim that $\hat{l}'$ is an $\widetilde{(\mathrm{RH})}$-vector. To show this, we exclude the possibilities 33.1 (i)–(iii), as in the proof of Lemma 24.2, and conclude that with $\hat{l}'$, $\hat{l}$ also comes from a Riemann surface. $\square$

LEMMA 33.3. *Let G be a finite group, and g_0 and N nonnegative integers. If $g_0 = 0$ then set, for $k \in G^\times$,*

$$q(k) = \max\left\{ \frac{-\widetilde{v}_\varphi}{\widetilde{T}_{k,\varphi}} \middle| \varphi \in \mathrm{Irr}(G), \widetilde{T}_{k,\varphi} \neq 0 \right\},$$

otherwise set $q(k) = 1$. Set $m = \max\{m(k) \mid k \in S(G)\}$. Suppose that for each $\widetilde{(\mathrm{RH})}$-vector with norm in the set $A = \{N, N+1, \ldots, N+m-1\}$, the corresponding character either comes from a Riemann surface or lies in *one of the infinite series of exceptions described in Section 23.2. Addition-ally suppose that no vector of norm in A fails to be an $\widetilde{(RH)}$-vector because of 33.1 (iii). Then each character corresponding to an $\widetilde{(RH)}$-vector $\hat{l}$ of norm at least N comes from a Riemann surface or lies in one of the above infinite series of exceptions, provided that $\hat{l}(h) \geq q(h) + m(h)$ for at least one $h \in G^\times$.*

Proof. Similarly as in the proof of Lemma 24.3, let $\hat{l} = (\hat{l}(k))_{k \in S(G)}$ be an $\widetilde{(RH)}$-vector of norm at least $N + m$ that does not lie in an infinite series of exceptions, and such that $\hat{l}(h) \geq q(h) + m(h)$. We have to show that the character corresponding to $\hat{l}$ comes from a Riemann surface, and this can be done, as in the proof of Lemma 24.3, by considering the vector $\hat{l}'$ that is obtained from $\hat{l}$ by decreasing $\hat{l}(h)$ by $m(h)$. $\qquad \square$

ALGORITHM 33.4. Let G be a group such that its matrix $\hat{T}$ is invertible, and g_0 a nonnegative integer. If at most finitely many $\widetilde{(RH)}$-vectors $\hat{l} = (\hat{l}(h))_{h \in S(G)}$ of G with $\hat{l}(1) = g_0$ do not come from Riemann surfaces and do not lie in the infinite series of exceptions described in Theorems 23.5, 23.7, and 23.8 then this algorithm computes the set V of these vectors.

1. If $g_0 = 1$ then compute the vectors $\hat{l}$ corresponding to the infinite series of exceptions, and set I to be the set of these vectors. For $g_0 \neq 1$, set $I = \emptyset$.

2. Compute the numbers $q(h)$ and $m(h)$ used in Lemma 33.3, for $h \in S(G)$. Set $N = \min\{q(h) \mid h \in S(G)\}$, $m = \max\{m(h) \mid h \in S(G)\}$, $V = \emptyset$, and $S = \emptyset$.

3. Loop over all those vectors $\hat{l}$ with $0 \leq \hat{l}(h) \leq q(h) + m(h) - 1$ for all $h \in S(G)$. Add those $\hat{l}$ to S that fail to be $\widetilde{(RH)}$-vectors because of 33.1 (iii). Add those $\hat{l}$ to V that are $\widetilde{(RH)}$-vectors, not contained in I, and such that the corresponding character does not come from a Riemann surface.

4. If $V \cup S$ contains no vector of norm N, $N+1$, ..., $N+m-1$ then return V and terminate the algorithm.

5. Compute the set $E \subseteq S(G)$ of those class representatives that are not con-tained in any proper normal subgroup of G. For each $h \in E$, compute the smallest value $b(h) \geq q(h)$ such that $\hat{M}(b(h), h)$ satisfies the assumptions of Lemma 33.2.

6. Increase N by 1. Loop over all those vectors $\hat{l}$ of norm $N+1$ with $\hat{l}(h) > q(h) + m(h) - 1$ for at least one $h \in S(G)$ and with $\hat{l}(h) < b(h)$ for all $h \in E$. Add those $\hat{l}$ to S that fail to be $\widetilde{(RH)}$-vectors because of 33.1 (iii). Add those $\hat{l}$ to V that are $\widetilde{(RH)}$-vectors, not contained in I, and such that the corresponding character does not come from a Riemann surface.

7. If $V \cup S$ contains no vector of norm N, $N+1$, ..., $N+m-1$ then return V and terminate the algorithm. Otherwise go to Step 6.

The validity of the algorithm is shown analogously to that of Algorithm 24.4. In computations with Algorithm 33.4 also, we can reduce the number of checks in Step 3 and Step 6 by storing a set of smallest representatives w.r.t. the partial ordering $\prec$ given by $\hat{l}' \prec \hat{l}$ if $\hat{l}'(h) \leq \hat{l}(h)$ and $\hat{l}'(h) \equiv \hat{l}(h) \pmod{m(h)}$ holds for all $h \in S(G)$; if h can be written as a product of $c(h) + 1$ elements in h^G, with $0 < c(h) < m(h)$, then the congruence can be weakened to $\hat{l}'(h) \equiv \hat{l}(h) \pmod{c(h)}$ in the case that $\hat{l}'(h) > 0$.

EXAMPLE 33.5 (cf. Example 24.6). We apply Algorithm 33.4 to the alternating group $G = A_4$ on 4 points. Table 24 shows the character table and the matrix $\hat{T}$ of G, with rows and columns arranged compatibly. We have

		1A	2A	3A	3B
1_G		1	1	1	1
χ_2		1	1	ζ_3	ζ_3^2
χ_3		1	1	ζ_3^2	ζ_3
χ_4		3	-1	0	0

$$\begin{pmatrix} 1 & 1 & 1 & 3 \\ 0 & 0 & 0 & 1 \\ 0 & 2/3 & 1/3 & 1 \\ 0 & 1/3 & 2/3 & 1 \end{pmatrix}$$

TABLE 24. Character Table and Matrix $\hat{T}$ for A_4

$m(2A) = 2$, $m(3A) = m(3B) = 3$. In the case $g_0 = 0$, we get $q(h) = 3$ for all $h \in G^\times$, so 180 vectors must be considered in Step 3; all characters of $\widetilde{(RH)}$-vectors in this set come from Riemann surfaces. Table 25 lists the smallest representatives w.r.t. the relation $\prec$ of G-characters that come from Riemann surfaces. We see that the algorithm is not terminated in Step 4, and compute $E = \{3A, 3B\}$, and $b(3A) = b(3A) = 4$. The algorithm is terminated in Step 7 with $N = 4$. The case $g_0 = 1$ is treated analogously, it turns out that all

g	g_0	$l(2A)$	$l(3A)$	$l(3B)$
9	0	0	1	4
5	0	0	2	2
13	0	0	3	3
9	0	0	4	1
4	0	1	0	3
12	0	1	1	4
8	0	1	2	2
4	0	1	3	0
12	0	1	4	1

g	g_0	$l(2A)$	$l(3A)$	$l(3B)$
3	0	2	1	1
13	0	0	6	0
13	0	0	0	6
13	1	0	0	3
9	1	0	1	1
13	1	0	3	0
4	1	1	0	0
12	1	1	1	1

TABLE 25. Representatives of $\prec$ for A_4

$\widetilde{(RH)}$-characters of G come from Riemann surfaces. ◇

34. Example: $G = L_2(q)$ for $q \equiv -1$ (mod 4)

In this section, let G be the projective special linear group $L_2(q)$, for $q \equiv -1$ (mod 4).

34.1. Finiteness Result. The irreducible characters of G can be found in [**Sch07**, p. 128], they are shown in Table 26; the parameter values are $1 \leq a, \alpha, \beta \leq \frac{q-3}{4}$ and $1 \leq b \leq \frac{q+1}{4}$.

	1	P	Q	A^a	B^b
1_G	1	1	1	1	1
φ_β	$q-1$	-1	-1	0	$-\zeta_{(q+1)/2}^{b\beta} - \zeta_{(q+1)/2}^{-b\beta}$
ψ	q	0	0	1	-1
μ_α	$q+1$	1	1	$\zeta_{(q-1)/2}^{a\alpha} + \zeta_{(q-1)/2}^{-a\alpha}$	0
θ_1	$\frac{q-1}{2}$	$\frac{-1+\sqrt{-q}}{2}$	$\frac{-1-\sqrt{-q}}{2}$	0	$-(-1)^b$
θ_2	$\frac{q-1}{2}$	$\frac{-1-\sqrt{-q}}{2}$	$\frac{-1+\sqrt{-q}}{2}$	0	$-(-1)^b$

TABLE 26. Irreducible Characters of $L_2(q)$ for $q \equiv -1$ (mod 4)

THEOREM 34.1. *Let* $G = L_2(q)$, *with* $q \equiv -1$ (mod 4), $q \geq 7$. *Then there are only finitely many G-characters that satisfy the condition* $\widetilde{(\mathrm{RH})}$ *but do not come from Riemann surfaces.*

Proof. G is simple for $q \geq 7$, so we want to apply Theorem 31.8. We have to show that the value $S(\sigma)$ introduced in the proof of Theorem 27.1 does not vanish for all $\sigma \in G^\times$. Note that only the classes of P and Q have nonreal character values.

$$S(P) = 1 + \frac{q-3}{4} \frac{-1}{q-1} + \frac{q-3}{4} \frac{1}{q+1}$$

$$+ \frac{2}{q-1} \left(\left(\frac{-1+\sqrt{-q}}{2} \right)^2 \frac{-1-\sqrt{-q}}{2} + \left(\frac{-1-\sqrt{-q}}{2} \right)^2 \frac{-1+\sqrt{-q}}{2} \right)$$

$$= 1 - \frac{q-3}{2(q^2-1)} - \frac{q+1}{2(q-1)} = \frac{q(q-3)}{2(q^2-1)} > 0$$

because $q > 3$. Reordering the summands yields $S(Q) = S(P)$.

$$S(A^a) = 1 + \frac{1}{q} + \frac{1}{q+1} \sum_{\alpha=1}^{(q-3)/4} \left(\zeta_{(q-1)/2}^{a\alpha} + \zeta_{(q-1)/2}^{-a\alpha} \right)^3$$

$$\geq 1 + \frac{1}{q} - \frac{4}{q+1} = \frac{(q-1)^2}{q(q+1)} > 0$$

by Lemma 27.2, and analogously

$$
\begin{aligned}
S(B^b) &= 1 - \frac{1}{q-1} \sum_{\beta=1}^{(q-3)/4} \left(\zeta_{(q+1)/2}^{b\beta} + \zeta_{(q+1)/2}^{-b\beta} \right)^3 - \frac{1}{q} + 2 \cdot \frac{-(-1)^b}{(q-1)/2} \\
&\geq 1 - \frac{1}{q-1} \left(\frac{q+1}{2} - 4 \cdot ((-1)^b + 1) \right) - \frac{1}{q} - \frac{4(-1)^b}{q-1} \\
&= \left(1 - \frac{q+1}{2(q-1)} \right) + \left(\frac{1}{q-1} - \frac{1}{q} \right) > 0
\end{aligned}
$$

because both summands are positive. $\qquad\square$

34.2. Exceptional Characters. In this section, we construct several $\widetilde{(\mathrm{RH})}$-vectors of $G = L_2(q)$, with $q > 3$, for which the corresponding characters do not come from Riemann surfaces. In order to show that these vectors are in fact $\widetilde{(\mathrm{RH})}$-vectors, we compute the rows of P, Q, and B^b in the matrix $\hat{T}$. Let $q = p^{2d+1}$, for a prime p; note that if $q \equiv -1$ (mod 4) then q cannot be a square, and also $p \equiv -1$ (mod 4).

LEMMA 34.2. *Let* $h \in P^G \cup Q^G$, *and* $\varphi \in \mathrm{Irr}(G) \setminus \{1_G, \theta_1, \theta_2\}$. *Then* $\widetilde{T}_{h,\varphi} = q(p-1)/(2p)$.

Proof. We have that h is of order p, so

$$
\widetilde{T}_{h,\varphi} = [\widetilde{\psi}_{G,h}, \varphi] = \sum_{i=1}^{p-1} \frac{i}{p} [(\varepsilon_h^{-i})^G, \varphi] = \sum_{i=1}^{p-1} \frac{i}{p} [\varepsilon_h^{-i}, \varphi_{\langle h \rangle}]
$$

holds. Since $(\varphi_\beta)_{\langle P \rangle} = \frac{q}{p} \cdot (\rho_{\langle P \rangle} - 1_{\langle P \rangle})$, $\psi_{\langle P \rangle} = \frac{q}{p} \cdot \rho_{\langle P \rangle}$, and $(\mu_\alpha)_{\langle P \rangle} = \frac{q}{p} \cdot (\rho_{\langle P \rangle} + 1_{\langle P \rangle})$. So $[\varepsilon_P^{-i}, \varphi_{\langle P \rangle}] = [\varepsilon_Q^{-i}, \varphi_{\langle P \rangle}] = q/p$ in all cases, and thus $\widetilde{T}_{h,\varphi} = (q/p) \cdot \sum_{i=1}^{p-1} i/p = q(p-1)/(2p)$, as claimed. $\qquad\square$

LEMMA 34.3. *Let* r_p, *resp.* n_p, *denote the sum of quadratic residues, resp. non-residues, modulo* p. *Then*

$$
\widetilde{T}_{P,\theta_1} = \widetilde{T}_{Q,\theta_2} = \frac{p^d(p-1)(p^d - 1)}{4} + p^{d-1} n_p
$$

and

$$
\widetilde{T}_{P,\theta_2} = \widetilde{T}_{Q,\theta_1} = \frac{p^d(p-1)(p^d - 1)}{4} + p^{d-1} r_p.
$$

Proof. First we compute $\theta_1 + \theta_2 = \frac{q}{p} \cdot \rho_{\langle P \rangle} - 1_{\langle P \rangle}$ and

$$
\widetilde{T}_{P,\theta_1} + \widetilde{T}_{P,\theta_2} = T_{\langle P \rangle, \theta_1 + \theta_2} = \frac{[\rho_{\langle P \rangle} - 1_{\langle P \rangle}, (\theta_1 + \theta_2)_{\langle P \rangle}]}{[\theta_1 + \theta_2, \theta_1 + \theta_2]} = \frac{q}{p} \cdot \frac{p-1}{2}.
$$

Because $n_p + r_p = \sum_{i=1}^{p-1} i = p(p-1)/2$ and

$$\frac{p^d(p-1)(p^d-1)}{2} + p^{d-1}\frac{p(p-1)}{2} = \frac{p^{2d}(p-1)}{2} = \frac{q(p-1)}{2p},$$

this means that it suffices to compute one of the above values. Note that the class of P consists of the inverses of the elements in the class of Q, so $\widetilde{T}_{P,\theta_1} = \widetilde{T}_{Q,\theta_2}$ is clear by Lemma 30.22 (a).

Let $res(p)$ denote the set of quadratic residues modulo p, $nres(p)$ the set of quadratic nonresidues modulo p, and $*u$, for $1 \le u \le p-1$, the field automorphism of $\mathbb{Q}(\zeta_p)$ that is defined by $\zeta_p^{*u} = \zeta_p^u$. By a theorem of Gauss, see for example [**Lan70**, pp. 85–90], we have

$$\sum_{j\in res(p)} \zeta_p^j = \sum_{j=1}^{(p-1)/2} \zeta_p^{j^2} = \frac{1}{2}(-1+\sqrt{-p})$$

in our situation that $p \equiv -1 \pmod 4$. As a consequence, $\sqrt{-p}^{*u} = \sqrt{-p}$ if and only if $u \in res(p)$. In other words, $\theta_1(P^j) = \frac{1}{2}(-1+\sqrt{-p})$ if $j \in res(p)$, and $\theta_1(P^j) = \frac{1}{2}(-1-\sqrt{-p})$ if $j \in nres(p)$. For $1 \le i \le p-1$, we thus compute

$$
\begin{aligned}
[\varepsilon_P^{-i}, \theta_1] &= \frac{1}{p}\sum_{j=0}^{p-1} \zeta_p^{ij} \cdot \theta_1(P^j) \\
&= \frac{1}{2p}\left((q-1) - \sum_{j=1}^{p-1}\zeta_p^{ij} + \sum_{j\in res(p)}\zeta_p^{ij}\sqrt{-q} - \sum_{j\in nres(p)}\zeta_p^{ij}\sqrt{-q}\right) \\
&= \frac{1}{2p}\left(q + 2\sqrt{-q}\sum_{j\in res(p)}\zeta_p^{ij} - \sqrt{-q}\sum_{j=1}^{p-1}\zeta_p^{ij}\right) \\
&= \frac{1}{2p}\left(q + 2\sqrt{-q}\left(\sum_{j\in res(p)}\zeta_p^j\right)^{*i} + \sqrt{-q}\right) \\
&= \frac{1}{2p}\left(q + 2\sqrt{-q}\sqrt{-p}^{*i}\right) \\
&= \begin{cases} (p^{2d}-p^d)/2, & i\in res(p), \\ (p^{2d}+p^d)/2, & i\in nres(p). \end{cases}
\end{aligned}
$$

From this, we get immediately

$$
\begin{aligned}
\widetilde{T}_{P,\theta_1} &= \sum_{i=1}^{p-1}\frac{i}{p}[\varepsilon_P^{-i}, \theta_1] = \frac{p-1}{2}\cdot\frac{p^{2d}-p^d}{2} + p^d\sum_{i\in nres(p)}\frac{i}{p} \\
&= \frac{p^d(p-1)(p^d-1)}{4} + p^{d-1}n_p.
\end{aligned}
$$

$\square$

REMARK 34.4. In the special case that $p = q$, we have $\widetilde{T}_{P,\theta_1} = \widetilde{T}_{Q,\theta_2} = n_p/p$ and $\widetilde{T}_{P,\theta_2} = \widetilde{T}_{Q,\theta_1} = r_p/p$.

LEMMA 34.5. Let $b \in \{1, 2, \ldots, (q+1)/4\}$, and $c = [\langle B \rangle : \langle B^b \rangle] \cdot (|B^b| - 1)$. For $\varphi \in \mathrm{Irr}(G) \setminus \{1_G, \theta_1, \theta_2\}$, we have $\widetilde{T}_{B^b,\varphi} = c$, except if $\varphi = \varphi_\beta$ such that $\frac{q+1}{2}$ does not divide $b\beta$; in this case, $\widetilde{T}_{B^b,\varphi_\beta} = c - 1$.

$\widetilde{T}_{B^b,\theta_1} = \widetilde{T}_{B^b,\theta_2} = \lfloor c/2 \rfloor$, the integral part of $c/2$.

Proof. Let $U = \langle B^b \rangle$, and $d = [\langle B \rangle : \langle B^b \rangle]$. The G-class of B^b is real, so $\widetilde{T}_{B^b,\varphi} = \frac{1}{2} \cdot [\rho_U - 1_U, \varphi_U]$. We compute

$$(\varphi_\beta)_U = 2d \cdot \rho_U - (\varepsilon_B^\beta)_U - (\varepsilon_B^{-\beta})_U, \quad \psi_U = 2d \cdot \rho_u - 1_U, \quad (\mu_\alpha)_U = 2d \cdot \rho_U,$$

and

$$(\theta_1)_U = (\theta_2)_U = d \cdot \rho_U - \chi,$$

where $\chi = 1_G$ if b is even, and χ is the nontrivial extension of $1_{\langle B^2 \rangle}$ to U if b is odd. Now the claim follows from $[\rho_U - 1_U, \rho_U] = |U| - 1$ and $c = d \cdot (|U| - 1)$. $\square$

We define the vectors $\hat{l}_0$, $\hat{l}_0^*$, $\hat{l}_{1,b}$, $\hat{l}_{1,b}^*$, and $\hat{l}_{2,b}$ by their nonzero values: $\hat{l}_0(P) = 2$ and $\hat{l}_0(Q) = 1$, $\hat{l}_0^*(P) = 1$ and $\hat{l}_0^*(Q) = 2$, $\hat{l}_{1,b}(P) = 2$ and $\hat{l}_{1,b}(B^b) = 1$, $\hat{l}_{1,b}^*(Q) = 2$ and $\hat{l}_{1,b}^*(B^b) = 1$, and $\hat{l}_{2,b}(P) = \hat{l}_{2,b}(Q) = \hat{l}_{2,b}(B^b) = 1$. Here the value b is chosen from the parameter range $\{1, 2, \ldots, \frac{q+1}{4}\}$, as in Lemma 34.5.

A vector $\hat{l}$ is an $\widetilde{(\mathrm{RH})}$-vector if the entries of $a = (\widetilde{v}_\varphi)_{\varphi \in P(G)} + \hat{l} \cdot \hat{T}$ are nonnegative integers and if the corresponding character has degree at least 2. First we show that for $\hat{l}$ equal to one of the above vectors, the entries of a are integers. This follows from the fact that the entries in the rows of P, Q, and B^b are integers, which is clear from the above calculations, except for $\widetilde{T}_{P,\theta_1}$ and $\widetilde{T}_{P,\theta_2}$. So we show that n_p/p and r_p/p are integers for $p > 3$.

LEMMA 34.6 (cf. [**Was96**, Exercise 4.5 on p. 46]). *Let p be a prime, $p > 3$, $p \equiv -1 \pmod 4$, $K = \mathbb{Q}(\sqrt{-p})$, and $h = h(K) = h^-(K)$ the (relative) class number of K. Then $h = (n_p - r_p)/p$.*

Proof. By [**Was96**, Theorem 4.17] (cf. Theorem C.8), we have $h(K) = Qw(-\frac{1}{2}B_{1,\chi})$. Here $Q = [E:WE^+] = 1$, because the groups E of units in K, W of roots of unity in K, and E^+ of units in $\mathbb{Q}$ are all equal to $\{\pm 1\}$. Furthermore $w = |W| = 2$, and χ is the unique quadratic character modulo p, with $\chi(a) = 1$ if $a \in res(p)$, and $\chi(a) = -1$ if $a \in nres(p)$. Thus

$$B_{1,\chi} = \frac{1}{p} \sum_{a=1}^{p-1} \chi(a) \cdot a = \frac{1}{p}(r_p - n_p).$$

□

Since both $n_p + r_p$ and $n_p - r_p$ are multiples of p, n_p and r_p themselves are also divisible by p.

REMARK 34.7. As a consequence, we conclude that for primes $p > 3$, $p \equiv -1$ (mod 4), the number

$$\sum_{u \in res(p)} \frac{\zeta_p^u}{1 - \zeta_p^u} = -\frac{r_p}{p} + \frac{n_p - r_p}{p} \cdot \frac{-1 + \sqrt{-p}}{2}$$

is an algebraic integer.

Now we show that $\hat{l}_0$ and $\hat{l}_0^*$ are $\widetilde{(RH)}$-vectors for $q = p$. The nonnegativity of the a_φ for $\hat{l}_0$ follows from the fact that

$$\tilde{v}_\varphi + (\hat{l}_0 \cdot \hat{T})_\varphi \geq -(p+1) + 3 \cdot \frac{p-1}{2} = \frac{p-5}{2} > 0$$

for $\varphi \in \mathrm{Irr}(G) \setminus \{1_G, \theta_1, \theta_2\}$,

$$\tilde{v}_{\theta_1} + (\hat{l} \cdot \hat{T})_{\theta_1} = -\frac{p-1}{2} + \frac{p-1}{2} + \frac{n_p}{p} > 0,$$

and

$$\tilde{v}_{\theta_2} + (\hat{l} \cdot \hat{T})_{\theta_2} = -\frac{p-1}{2} + \frac{p-1}{2} + \frac{r_p}{p} > 0.$$

The argument for $\hat{l}_0^*$ is the same, with the roles of P and Q interchanged.

The following lemma shows that the characters corresponding to $\hat{l}_0$ and $\hat{l}_0^*$ cannot come from Riemann surfaces.

LEMMA 34.8. Let p be a prime, $p \equiv -1$ (mod 4), $G = L_2(p)$, and P, Q as in Table 26 on page 164. Then $\mathrm{Epi}_{(P^G, P^G, Q^G)}(0, G) = \mathrm{Epi}_{(P^G, Q^G, Q^G)}(0, G) = \emptyset$.

Proof. The class P^G consists of the inverses of the elements in Q^G, so Theorem 14.1 yields that

$$|\mathrm{Hom}_{(P^G, P^G, Q^G)}(0, G)| = \frac{1}{|G|} \sum_{\chi \in \mathrm{Irr}(G)} |P^G|^3 \cdot \frac{\chi(P)^2 \chi(Q)}{\chi(1)}$$

$$= \frac{|P^G|^3}{|G|} \cdot S(P) = \frac{(p^2-1)(p-3)}{8},$$

since

$$|G| = p(p^2-1)/2, \quad |P^G| = |G|/p, \quad \text{and} \quad S(P) = \frac{p(p-3)}{2(p^2-1)},$$

see the proof of Theorem 34.1. Repeating the argument with the roles of P and Q interchanged, we get $|\mathrm{Hom}_{(P^G, Q^G, P^G \cup Q^G)}(0, G)| = (p^2-1)(p-3)/4$.

Each subgroup U of order p in G contains exactly $(p-1)/2$ elements in each of the classes P^G, Q^G. Choosing any element $\sigma \in U \cap P^G$ and any element except σ^{-1} in $U \cap Q^G$ gives us an element in $\mathrm{Hom}_{(P^G, Q^G, P^G \cup Q^G)}(0, G)$, so exactly $\frac{p-1}{2} \cdot \frac{p-3}{2}$ such homomorphisms have image U. There are $p+1$ conjugates of U in G, hence $\mathrm{Epi}_{(P^G, Q^G, P^G \cup Q^G)}(0, G) = \emptyset$. □

Hence we obtain the following.

THEOREM 34.9. Let $\chi = \sum_{\varphi \in \mathrm{Irr}(G)} a_\varphi \varphi$, with

$$a_{1_G} = 0, \quad a_{\varphi_\beta} = (p-1)/2, \quad a_\psi = (p-3)/2,$$
$$a_{\mu_\alpha} = (p-5)/2, \quad a_{\theta_1} = n_p/p, \quad a_{\theta_2} = r_p/p.$$

Then $\hat{l}_\chi$ is the vector $\hat{l}_0$ defined above, and χ is an $\widetilde{(\mathrm{RH})}$-character of degree $1 + (p-3)(p^2-1)/4$ of G that does not come from a Riemann surface.

Proof. We have computed above that

$$a_\varphi = -\varphi(1) + \hat{l}_0(P) \cdot \widetilde{T}_{P,\varphi} + \hat{l}_0(Q) \cdot \widetilde{T}_{Q,\varphi}$$

holds for all $\varphi \in \mathrm{Irr}(G)$; the degree of χ can then be derived easily. □

Next we consider the other vectors defined above. The entries of $(\widetilde{v}_\varphi)_{\varphi \in P(G)} + \hat{l} \cdot \widetilde{T}$ are integers for $\hat{l} \in \{\hat{l}_{1,b}, \hat{l}_{1,b}^*, \hat{l}_{2,b}\}$, with arbitrary b, since the relevant $\widetilde{T}_{h,\varphi}$ are integers. The nonnegativity is characterized in the following lemma.

LEMMA 34.10. Let b and c be as in Lemma 34.5, and $q = p^{2d+1} > 3$. Then the following hold:

(i) $\hat{l}_{1,b}$ and $\hat{l}_{1,b}^*$ are $\widetilde{(\mathrm{RH})}$-vectors if and only if one of the conditions
  (1) $3 < p < q$,
  (2) $3 < p = q$ and $c/2 \geq h(\mathbb{Q}(\sqrt{-p}))$,
  (3) $p = 3$ and $2 \cdot \lfloor c/2 \rfloor \geq 3^{2d} + 2 \cdot 3^{d-1} - 1$
  holds;
(ii) $\hat{l}_{2,b}$ is an $\widetilde{(\mathrm{RH})}$-vector if and only if one of the conditions
  (4) $3 < p$,
  (5) $p = 3$ and $c \geq q/3 + 1$
  holds.

Proof. First we note that in general,

(∗) $c = |B| - |B|/|B^b| \geq |B| - |B|/2 = (q+1)/4$

holds.

(i) For $\varphi \in \mathrm{Irr}(G) \setminus \{1_G, \theta_1, \theta_2\}$,

$$\tilde{v}_\varphi + (\hat{l}_{1,b} \cdot \hat{T})_\varphi = -\varphi(1) + \tilde{T}_{B^b,\varphi} + 2 \cdot \tilde{T}_{P,\varphi}$$

$$\geq -q - 1 + c + \frac{q(p-1)}{p} = c - \left(\frac{q}{p} + 1\right),$$

so condition (3) is sufficient for $p = 3$. If $p \geq 7$ then $(*)$ implies that

$$c - \left(\frac{q}{p} + 1\right) \geq \frac{q+1}{4} - \frac{q}{p} - 1 = \frac{(p-4)(q-3) - 12}{4p} \geq 0.$$

Because $n_p > r_p$, we have $\tilde{T}_{P,\theta_1} > \tilde{T}_{P,\theta_2}$ and thus $\tilde{v}_{\theta_1} + (\hat{l}_{1,b} \cdot \hat{T})_{\theta_1} > \tilde{v}_{\theta_2} + (\hat{l}_{1,b} \cdot \hat{T})_{\theta_2}$, so it suffices to look at the value for θ_2. If $p = q$ then

$$\tilde{v}_{\theta_2} + (\hat{l}_{1,b} \cdot \hat{T})_{\theta_2} = -\frac{p-1}{2} + \frac{2r_p}{p} + \left\lfloor \frac{c}{2} \right\rfloor = \left\lfloor \frac{c}{2} \right\rfloor - h(\mathbb{Q}(\sqrt{-p}))$$

is nonnegative by condition (2). If $3 \neq p < q$ then we use $(*)$ and get

$$\tilde{v}_{\theta_2} + (\hat{l}_{1,b} \cdot \hat{T})_{\theta_2} = -\frac{p^{2d+1} - 1}{2} + \left\lfloor \frac{c}{2} \right\rfloor + \frac{p^d(p-1)(p^d-1)}{2} + 2p^{d-1}r_p$$

$$\geq -\frac{p^{2d+1} - 1}{2} + \frac{p^{2d+1} - 3}{8} + \frac{p^{2d+1} - p^{2d} - p^{d+1} + p^d}{2}$$

$$= \frac{1}{8}(p^{2d+1} + 1 - 4p^{2d} - 4p^{d+1} + 4p^d)$$

$$> \frac{p^d}{8}(p^{d+1} - 4p^d - 4p + 4)$$

$$= \frac{p^d}{8}((p^d - 4)(p - 4) - 12) > 0.$$

Finally, if $p = 3$ then $\tilde{v}_{\theta_2} + (\hat{l}_{1,b} \cdot \hat{T})_{\theta_2}$ is nonnegative if and only if condition (3) holds; note that $r_3 = 1$.

(ii) For $\varphi \in \mathrm{Irr}(G) \setminus \{1_G, \theta_1, \theta_2\}$,

$$\tilde{v}_\varphi + (\hat{l}_{2,b} \cdot \hat{T})_\varphi = -\varphi(1) + \tilde{T}_{B^b,\varphi} + \tilde{T}_{P,\varphi} + \tilde{T}_{Q,\varphi}$$

$$\geq -q - 1 + c + \frac{q(p-1)}{p} = c - \left(\frac{q}{p} + 1\right)$$

is nonnegative by condition (4) or (5), as we computed in (i). In the remaining cases,

$$\tilde{v}_{\theta_1} + (\hat{l}_{2,b} \cdot \hat{T})_{\theta_1} = \tilde{v}_{\theta_2} + (\hat{l}_{2,b} \cdot \hat{T})_{\theta_2} = -\frac{q-1}{2} + \frac{q(p-1)}{2p} + \left\lfloor \frac{c}{2} \right\rfloor$$

$$\geq \frac{1}{2}\left(1 - \frac{q}{p} + c - 1\right) = \frac{1}{2}\left(c - \frac{q}{p}\right)$$

is nonnegative for $p = 3$, by condition (5); if $p > 3$ then $(*)$ yields that

$$\frac{1}{2}\left(c - \frac{q}{p}\right) \geq \frac{1}{2}\left(\frac{q+1}{4} - \frac{q}{p}\right) = \frac{pq + p - 4q}{8p}$$

$$= \frac{(p-4)(q+1) + 4}{8p} > 0.$$

$\square$

LEMMA 34.11. *If b is odd then* $\mathrm{Hom}_{((B^b)^G, P^G, P^G)}(0, G) = \emptyset$. *If b is even then* $\mathrm{Hom}_{((B^b)^G, P^G, Q^G)}(0, G) = \emptyset$.

Proof. Before we apply Theorem 14.1, we note that

$$\sum_{\beta=1}^{(q-3)/4} \left(-\zeta_{(q+1)/2}^{b\beta} - \zeta_{(q+1)/2}^{-b\beta}\right) = -\sum_{\beta=1}^{(q-1)/2} \zeta_{(q+1)/2}^{b\beta} + \zeta_{(q+1)/2}^{b(q+1)/4} = 1 + (-1)^b.$$

Using this equality, we find that

$$\sum_{\chi \in \mathrm{Irr}(G)} \frac{\chi(B^b)\chi(P)^2}{\chi(1)} = 1 + \frac{1}{q-1}\sum_{\beta=1}^{(q-3)/4}\left(-\zeta_{(q+1)/2}^{b\beta} - \zeta_{(q+1)/2}^{-b\beta}\right)$$

$$- \frac{2(-1)^b}{q-1}\cdot\left(\left(\frac{-1+\sqrt{-q}}{2}\right)^2 + \left(\frac{-1-\sqrt{-q}}{2}\right)^2\right)$$

$$= 1 + \frac{1 + (-1)^b}{q-1} + (-1)^b$$

$$= (1 + (-1)^b)\frac{q}{q-1}$$

is zero if b is odd, and

$$\sum_{\chi \in \mathrm{Irr}(G)} \frac{\chi(B^b)\chi(P)\chi(Q)}{\chi(1)} = 1 + \frac{1}{q-1}\sum_{\beta=1}^{(q-3)/4}(-\zeta_{(q+1)/2}^{b\beta} - \zeta_{(q+1)/2}^{-b\beta})$$

$$- \frac{4(-1)^b}{q-1}\cdot\left(\frac{-1+\sqrt{-q}}{2}\right)\left(\frac{-1-\sqrt{-q}}{2}\right)$$

$$= 1 + \frac{1 + (-1)^b}{q-1} - \frac{q+1}{q-1}(-1)^b$$

$$= (1 - (-1)^b)\frac{q}{q-1}$$

vanishes if b is even. $\square$

The following consequence is obvious.

COROLLARY 34.12. *If b is odd then the G-class functions corresponding to $\hat{l}_{1,b}$, $\hat{l}_{1,b}^*$ do not come from Riemann surfaces. If b is even then the G-class function corresponding to $\hat{l}_{2,b}$ does not come from a Riemann surface.*

Together with Lemma 34.10, this gives us further examples of $\widetilde{(\text{RH})}$-characters of G that do not come from Riemann surfaces. In particular, we see that the condition $\widetilde{(\text{RH})}$ for a G-character χ does *not* guarantee that $\text{Hom}_{L(\chi)}(\Gamma(\chi), G)$ is nonempty. So it may for example happen that Lemma 22.11 or Lemma 30.16 is not applicable because there is no $\Phi \in \text{Hom}_C(\Gamma(\chi), G)$.

34.3. Small Values of q. We look at $L_2(q)$ for $q \in \{3, 7, 11, 19, 23, 27, 31\}$. The group $L_2(3) \cong A_4$ has been considered in Example 33.5.

For $L_2(7)$, the order of B is 4, so we get the exceptional vectors $\hat{l}_0$ and $\hat{l}_0^*$ for genus 49, $\hat{l}_{1,1}$ and $\hat{l}_{1,1}^*$ for genus 40, and $\hat{l}_{2,2}$ for genus 19. Additionally, there is one $\widetilde{(\text{RH})}$-vector l, given by the nonzero values $l(1\text{A}) = l(2\text{A}) = 1$, whose character is of degree 43 and does not come from a Riemann surface. Thus exactly six $\widetilde{(\text{RH})}$-characters of $L_2(7)$ do not come from Riemann surfaces. This confirms [**KK91**, Theorem on p. 349]; note that $L_2(7) \cong L_3(2) \cong GL_3(2)$, and that the condition (CY) used in [**KK91**] is equivalent to our condition $\widetilde{(\text{RH})}$ in this case, since the nonreal elements of $L_2(7)$ have prime order.

In the remaining small examples, all $\widetilde{(\text{RH})}$-characters that do not come from Riemann surfaces are described by the vectors studied in Section 34.2, they are listed in Table 27.

q	$\lvert B \rvert$	g	$\hat{l}$	q	$\lvert B \rvert$	g	$\hat{l}$	q	$\lvert B \rvert$	g	$\hat{l}$
11	6	106	$\hat{l}_{1,3}, \hat{l}_{1,3}^*$	23	12	1 255	$\hat{l}_{2,6}$	31	16	3 241	$\hat{l}_{2,8}$
		161	$\hat{l}_{2,2}$			1 761	$\hat{l}_{2,4}$			5 101	$\hat{l}_{2,4}$
		216	$\hat{l}_{1,1}, \hat{l}_{1,1}^*$			2 014	$\hat{l}_{1,3}, \hat{l}_{1,3}^*$			6 031	$\hat{l}_{2,2}$
		241	$\hat{l}_0, \hat{l}_0^*$			2 267	$\hat{l}_{2,2}$			6 496	$\hat{l}_{1,1}, \hat{l}_{1,1}^*$
19	10	676	$\hat{l}_{1,5}, \hat{l}_{1,5}^*$			2 520	$\hat{l}_{1,1}, \hat{l}_{1,1}^*$			6 721	$\hat{l}_0, \hat{l}_0^*$
		1 189	$\hat{l}_{2,2}$			2 641	$\hat{l}_0, \hat{l}_0^*$				
		1 360	$\hat{l}_{1,1}, \hat{l}_{1,1}^*$	27	14	937	$\hat{l}_{2,2}$				
		1 441	$\hat{l}_0, \hat{l}_0^*$			1 288	$\hat{l}_{1,1}, \hat{l}_{1,1}^*$				

TABLE 27. Exceptional Vectors for $L_2(q)$ with Small $q \equiv -1 \pmod 4$

Note that for $L_2(27)$, $\hat{l}_0$ and $\hat{l}_0^*$ do not occur because 27 is not a prime, and $\hat{l}_{1,7}$ and $\hat{l}_{1,7}^*$ do not occur because for $q = 27$ and $b = 7$, $c = 7 < q/3 + 1$, see Lemma 34.10.

35. Example: $G = A_n$

In this section, let G be the alternating group A_n.

35.1. Finiteness Result. By Lemma 29.2, we have $\widetilde{rep}(\sigma^G) = 1$ for $G = A_n$ with $n \geq 5$, so Theorem 31.8 yields immediately

THEOREM 35.1. *Let $G = A_n$ with $n \geq 5$. There are only finitely many $\widetilde{(RH)}$-characters of G that do not come from Riemann surfaces.*

35.2. Small Values of n. The group A_3, cyclic of order 3, has been considered in Section 30.2, A_4 in Example 33.5, and the ambivalent groups A_5 and A_6 in Sections 27.3 and 28.2, respectively.

Exactly the following two vectors $\hat{l}_1$ and $\hat{l}_2$ correspond to $\widetilde{(RH)}$-characters of A_7 that do not come from Riemann surfaces; we list only the nonzero values.

$$\hat{l}_1(3\mathsf{B}) = 2, \quad \hat{l}_1(7\mathsf{A}) = 1 \quad \text{and} \quad \hat{l}_2(3\mathsf{B}) = 2, \quad \hat{l}_2(7\mathsf{B}) = 1.$$

36. Example: Suzuki Groups $Sz(q)$

In this section, let G be the Suzuki group $Sz(q)$, for q an odd power of 2.

36.1. Finiteness Result. Table 28 shows the character table of $Sz(q)$, as it is given in [**Bur79**]. We define r by $r^2 = 2q$, and set

$$\varepsilon_d = \zeta^d_{q+r+1} + \zeta^{-d}_{q+r+1} + \zeta^{qd}_{q+r+1} + \zeta^{-qd}_{q+r+1},$$
$$\delta_e = \zeta^e_{q-r+1} + \zeta^{-e}_{q-r+1} + \zeta^{qe}_{q-r+1} + \zeta^{-qe}_{q-r+1}.$$

The parameters a and s run over the integers from 1 to $(q-2)/2$, b and l run over any set of numbers from 1 to $q+r$ where exactly one number from the set $\{x, -x \pmod{q+r+1}, qx \pmod{q+r+1}, -qx \pmod{q+r+1}\}$ is taken, and c and u run over any set of numbers from 1 to $q-r$ where exactly one number from the set $\{x, -x \pmod{q-r+1}, qx \pmod{q-r+1}, -qx \pmod{q-r+1}\}$ is taken.

THEOREM 36.1. *Let $G = Sz(q)$, with $q > 2$. Then there are only finitely many $\widetilde{(RH)}$-characters of G that do not come from Riemann surfaces.*

Proof. G is simple for $q \geq 8$. We show that $\widetilde{rep}(\sigma^G) < \infty$ for all $\sigma \in G^\times$, then the claim follows from Theorem 31.8.

The elements x, y, and z have orders $q-1$, $q+r+1$, and $q-r+1$, respectively, in particular these orders are odd. Since each of x^a, y^b, z^c is G-conjugate to its inverse, we have $\widetilde{rep}(\sigma^G) \leq |\sigma|$ for σ in these classes, see Remark 23.9.

	1	x^a	y^b	z^c	t	f	f^{-1}
1_G	1	1	1	1	1	1	1
Π	q^2	1	-1	-1	0	0	0
Γ_1	$r(q-1)/2$	0	1	-1	$-r/2$	$r\zeta_4/2$	$-r\zeta_4/2$
Γ_2	$r(q-1)/2$	0	1	-1	$-r/2$	$-r\zeta_4/2$	$r\zeta_4/2$
Ω_s	q^2+1	$\zeta_{q-1}^{sa}+\zeta_{q-1}^{-sa}$	0	0	1	1	1
Θ_l	$(q-1)(q-r+1)$	0	$-\varepsilon_{lb}$	0	$r-1$	-1	-1
Λ_u	$(q-1)(q+r+1)$	0	0	$-\delta_{uc}$	$-r-1$	-1	-1

TABLE 28. Irreducible Characters of $Sz(q)$

So we compute $\widetilde{rep}(\sigma^G) = 2$ for $\sigma = t$ and $\sigma = f$, by showing that $S(\sigma)$ does not vanish.

$$
\begin{aligned}
S(f) &= 1 + \frac{1}{r(q-1)/2}\left(\left(\frac{r\zeta_4}{2}\right)^2 \cdot \frac{-r\zeta_4}{2} + \left(\frac{-r\zeta_4}{2}\right)^2 \cdot \frac{r\zeta_4}{2}\right) + \frac{(q-2)/2}{q^2+1} \\
&\quad - \frac{(q+r)/4}{(q-1)(q-r+1)} - \frac{(q-r)/4}{(q-1)(q+r+1)} \\
&= 1 + \frac{(q-2)/2}{q^2+1} - \frac{(q+r)(q+r+1)+(q-r)(q-r+1)}{4(q-1)(q^2+1)} \\
&= \frac{(q-2)q(q+1)}{(q-1)(q^2+1)} > 0
\end{aligned}
$$

for $q > 2$, $S(f^{-1}) = S(f)$, and

$$
\begin{aligned}
S(t) &= 1 - 2\frac{(r/2)^3}{r(q-1)/2} + \frac{(q-2)/2}{q^2+1} + \frac{q+r}{4} \cdot \frac{(r-1)^3}{(q-1)(q-r+1)} \\
&\quad + \frac{q-r}{4} \cdot \frac{(-r-1)^3}{(q-1)(q+r+1)} \\
&= \frac{(q-2)q^2}{(q-1)(q^2+1)} > 0
\end{aligned}
$$

for $q > 2$. □

36.2. Small Values of q. The group $Sz(2)$ is isomorphic with $5:4$, the Frobenius group of order 20; all its $\widetilde{(RH)}$-characters come from Riemann surfaces. All $\widetilde{(RH)}$-characters of $Sz(8)$ and $Sz(32)$ also come from Riemann surfaces.

37. Example: Sporadic Simple Groups

37.1. Finiteness Result.

THEOREM 37.1. *Only finitely many* $(\widetilde{RH})$-*characters of sporadic simple groups do not come from Riemann surfaces.*

Proof. Let G be a sporadic simple group. In order to apply Theorem 31.8, we compute the values $\widetilde{rep}(\sigma^G)$ for all $\sigma \in G^\times$, using the character table of G available in GAP. It turns out that the $rep_{\max}$ value in Table 16 on page 110 is in fact the maximal $\widetilde{rep}(\sigma^G)$ value for G. □

37.2. The Mathieu Group M_{11}.

Exactly the following six vectors $\hat{l}$ of M_{11} correspond to $(\widetilde{RH})$-characters that do not come from Riemann surfaces.

g	g_0	2A	3A	4A	5A	6A	8A	8B	11A	11B
1 057	0	0	1	0	2	0	0	0	0	0
961	0	0	2	0	0	0	0	0	0	1
961	0	0	2	0	0	0	0	0	1	0
1 261	0	1	0	0	0	0	0	0	1	1
991	0	1	0	0	0	0	1	1	0	0
1 981	1	1	0	0	0	0	0	0	0	0

37.3. The Janko Group J_1.

For the simple group J_1 of order 175 560, all $(\widetilde{RH})$-characters come from Riemann surfaces.

APPENDIX A

Abelian Invariants

This appendix collects some well-known statements about abelian groups, which are needed to describe the commutator factor groups of Fuchsian groups.

Each finitely generated abelian group is isomorphic to an additive group

$$\mathbb{Z}^R \oplus \bigoplus_{p \in P} \bigoplus_{i=1}^{s_p} (\mathbb{Z}/p^i\mathbb{Z})^{N_{p,i}}$$

for a set P of primes. The numbers R and $N_{p,i}$ are unique if we require $1 \le N_{p,1} \le N_{p,2} \le \cdots \le N_{p,s_p}$. They are called the *abelian invariants* of the group.

Note that the abelian invariants do in general not immediately describe the decomposition into a direct sum of maximal cyclic subgroups. For example, a cyclic group of order 6 has abelian invariants 2 and 3.

LEMMA A.1. *Let p be a prime, $A \cong \mathbb{Z}^R \oplus \bigoplus_{i=1}^s (\mathbb{Z}/p^i\mathbb{Z})^{N_i}$, and $B \cong \mathbb{Z}^r \oplus \bigoplus_{i=1}^s (\mathbb{Z}/p^i\mathbb{Z})^{n_i}$. There is an epimorphism $A \to B$ if and only if*

$$(*) \qquad r + \sum_{i=j}^s n_i \le R + \sum_{i=j}^s N_i \quad \text{for} \quad 1 \le j \le s+1.$$

Proof. Condition $(*)$ is sufficient because we can map each of the first r generators of A to the generators of infinite order in B, each of the next n_s generators of A to generators of order p^s in B, and so on.

For the converse, we define $P_i(A) = \{\sigma^{p^i} \mid \sigma \in A\}$ for $i \ge 0$ and any abelian group A. Then $P_i(A)$ is a subgroup of A, and for each homomorphism Φ with domain A, we have $\Phi(P_i(A)) = P_i(\Phi(A))$. This means that the given epimorphism $\Phi\colon A \to B$ induces an epimorphism $P_i(A)/P_{i+1}(A) \to P_i(B)/P_{i+1}(B)$. The subquotients $P_i(A)/P_{i+1}(A)$ and $P_i(B)/P_{i+1}(B)$ are elementary abelian p-groups of ranks $R + \sum_{j=i}^s N_j$ and $r + \sum_{j=i}^s n_j$, respectively. $\square$

LEMMA A.2. *Let $A \cong \mathbb{Z}^R \oplus T_A$ and $B \cong \mathbb{Z}^r \oplus T_B$ be (additive) abelian groups, with T_A and T_B finite. There is an epimorphism $\Phi\colon A \to B$ if and only if, for each prime p that divides $|T_B|$, there is an epimorphism $\Phi_p\colon \mathbb{Z}^R \oplus Syl_p(T_A) \to \mathbb{Z}^r \oplus Syl_p(T_B)$, where Syl_p denotes the Sylow p-subgroup.*

Proof. If the epimorphism Φ is given then clearly it maps the torsion subgroup T_A of A into the torsion subgroup T_B of B. Define the group Q by $Q \oplus Syl_p(T_A) = T_A$. Then $\Phi(Q)$ is a subgroup of T_B of order coprime to p, hence the image of the restriction of Φ to $\mathbb{Z}^R \oplus Syl_p(T_A)$ contains $\mathbb{Z}^r \oplus Syl_p(T_B)$. So the composition of this restriction with the canonical projection $\mathbb{Z}^r \oplus T_B \to \mathbb{Z}^r \oplus Syl_p(T_B)$ yields Φ_p.

Conversely, we choose the epimorphisms Φ_p as in the proof of Lemma A.1, and define Φ by mapping the first r generators of infinite order in A to generators of infinite order in B, mapping the remaining generators of infinite order to the sum of their images under the Φ_p, and mapping the p-torsion as Φ_p does. $\qquad\square$

The abelian invariants of a group $\Gamma = \Gamma(g_0; m_1, m_2, \dots, m_r)$ with presentation

$$\left\langle a_1, b_1, a_2, b_2, \dots, a_g, b_g, c_1, c_2, \dots, c_r \,\middle|\, c_1^{m_1}, c_2^{m_2}, \dots, c_r^{m_r}, \prod_{i=1}^{g}[a_i, b_i] \prod_{j=1}^{r} c_j \right\rangle$$

as in Theorem 3.2 can be computed easily from this presentation.

LEMMA A.3. *Let* $\Gamma = \Gamma(g_0; m_1, m_2, \dots, m_r)$, *and* P *be the set of all primes dividing some of the* m_i. *For* $p \in P$, *let the p-parts of the periods* m_i *be* $p^{\mu_1(p)}$, $p^{\mu_2(p)}, \dots, p^{\mu_r(p)}$, *ordered so that* $0 \le \mu_1(p) \le \mu_2(p) \le \cdots \le \mu_r(p)$. *Then*

$$\Gamma/\Gamma' \cong \mathbb{Z}^{2g_0} \oplus \bigoplus_{p \in P} \bigoplus_{i=1}^{r-1} (\mathbb{Z}/p^{\mu_i(p)}\mathbb{Z}).$$

Proof. Set

$$M = \begin{pmatrix} m_1 & & & \\ & m_2 & & \\ & & \ddots & \\ & & & m_r \\ 1 & 1 & \cdots & 1 \end{pmatrix},$$

and let $\langle M \rangle$ denote the $\mathbb{Z}$-span of the rows of M. We have $\Gamma/\Gamma' \cong \mathbb{Z}^{2g_0} \oplus \mathbb{Z}^r/\langle M \rangle$, via the isomorphism that maps the hyperbolic generators of Γ to linearly independent generators of the free abelian group $\mathbb{Z}^{2g_0}$, and the i-th elliptic generator to the coset of the i-th standard basis vector of $\mathbb{Z}^r$ by $\langle M \rangle$.

$\mathbb{Z}^r/\langle M\rangle$ is a finite group. Its Sylow p-subgroup is isomorphic to $\mathbb{Z}^r/\langle M(p)\rangle$, with

$$M(p) = \begin{pmatrix} p^{\mu_1(p)} & & & & \\ & p^{\mu_2(p)} & & & \\ & & \ddots & & \\ & & & p^{\mu_r(p)} & \\ 1 & 1 & \cdots & 1 \end{pmatrix}.$$

This can be seen for example by considering, for $1 \leq i \leq r$, the rows $[0, 0, \ldots, 0, p^{\mu_i(p)}, 0, \ldots, 0]$, which lie in the kernel of the projection from $\mathbb{Z}^r/\langle M\rangle$ onto its largest p-factor.

Because

$$\mathbb{Z}^r/\langle M(p)\rangle \;=\; \mathbb{Z}^r \Bigg/ \Bigg\langle \begin{pmatrix} p^{\mu_1(p)} & & & & \\ & p^{\mu_2(p)} & & & \\ & & \ddots & & \\ & & & p^{\mu_{r-1}(p)} & \\ -p^{\mu_r(p)} & -p^{\mu_r(p)} & \cdots & -p^{\mu_r(p)} & 0 \\ 0 & 0 & \cdots & 0 & 1 \end{pmatrix} \Bigg\rangle$$

$$=\; \mathbb{Z}^r \Bigg/ \Bigg\langle \begin{pmatrix} p^{\mu_1(p)} & & & \\ & p^{\mu_2(p)} & & \\ & & \ddots & \\ & & & p^{\mu_{r-1}(p)} \\ & & & 1 \end{pmatrix} \Bigg\rangle,$$

we have $\mathbb{Z}^r/\langle M(p)\rangle \cong (\mathbb{Z}/p^{\mu_1(p)}\mathbb{Z}) \oplus (\mathbb{Z}/p^{\mu_2(p)}\mathbb{Z}) \oplus \cdots \oplus (\mathbb{Z}/p^{\mu_{r-1}(p)}\mathbb{Z})$.   □

APPENDIX B

Irreducible Characters

We list the irreducible representations that come from Riemann surfaces, see Section 19. The table is sorted according to genus g and the order of the group G. The third column contains the period vectors $(m_1, m_2, \ldots, m_r)$, the fourth column a description of the isomorphism type of G, and the last column representing matrices for the first $r - 1$ generators. The values b_7 and b_7^* that occur in the 3-dimensional representation of $L_3(2)$ are defined on page 43.

g	$\lvert G \rvert$	Periods	G	$\rho(c_1), \rho(c_2), \ldots, \rho(c_{r-1})$
2	6	$(2,2,3,3)$	D_6	$\begin{pmatrix} . & 1 \\ 1 & . \end{pmatrix}, \begin{pmatrix} . & 1 \\ 1 & . \end{pmatrix}, \begin{pmatrix} \zeta_3 & . \\ . & \zeta_3^2 \end{pmatrix}$
	8	$(4,4,4)$	Q_8	$\begin{pmatrix} \zeta_4 & . \\ . & \zeta_4^3 \end{pmatrix}, \begin{pmatrix} . & -1 \\ 1 & . \end{pmatrix}$
		$(2,2,2,4)$	D_8	$\begin{pmatrix} -1 & . \\ . & -1 \end{pmatrix}, \begin{pmatrix} . & 1 \\ . & -1 \end{pmatrix}, \begin{pmatrix} . & 1 \\ 1 & . \end{pmatrix}$
	12	$(3,4,4)$	6.2	$\begin{pmatrix} \zeta_3 & . \\ . & \zeta_3^2 \end{pmatrix}, \begin{pmatrix} . & -1 \\ 1 & . \end{pmatrix}$
		$(2,2,2,3)$	D_{12}	$\begin{pmatrix} -1 & . \\ . & -1 \end{pmatrix}, \begin{pmatrix} . & 1 \\ 1 & . \end{pmatrix}, \begin{pmatrix} . & \zeta_6^5 \\ \zeta_6 & . \end{pmatrix}$
	16	$(2,4,8)$	QD_{16}	$\begin{pmatrix} . & 1 \\ 1 & . \end{pmatrix}, \begin{pmatrix} . & \zeta_8^5 \\ \zeta_8^7 & . \end{pmatrix}$
	24	$(3,3,4)$	$SL_2(3)$	$\frac{1}{2}\begin{pmatrix} -1+\zeta_4 & 1+\zeta_4 \\ -1+\zeta_4 & -1-\zeta_4 \end{pmatrix}, \frac{1}{2}\begin{pmatrix} -1+\zeta_4 & 1-\zeta_4 \\ -1-\zeta_4 & -1-\zeta_4 \end{pmatrix}$
		$(2,4,6)$	$(2 \times 6).2$	$\begin{pmatrix} . & 1 \\ 1 & . \end{pmatrix}, \begin{pmatrix} . & \zeta_6^5 \\ \zeta_3^2 & . \end{pmatrix}$
	48	$(2,3,8)$	$GL_2(3)$	$\begin{pmatrix} . & \zeta_8 \\ \zeta_8^7 & . \end{pmatrix}, \frac{1}{2}\begin{pmatrix} -1+\zeta_4 & 1-\zeta_4 \\ -1-\zeta_4 & -1-\zeta_4 \end{pmatrix}$
3	12	$(2,2,3,3)$	A_4	$\begin{pmatrix} 1 & . & . \\ . & -1 & . \\ . & . & -1 \end{pmatrix}, \begin{pmatrix} 1 & . & . \\ . & -1 & . \\ . & . & -1 \end{pmatrix}, \begin{pmatrix} . & . & 1 \\ 1 & . & . \\ . & 1 & . \end{pmatrix}$
	21	$(3,3,7)$	$7:3$	$\begin{pmatrix} . & . & 1 \\ 1 & . & . \\ . & 1 & . \end{pmatrix}, \begin{pmatrix} . & \zeta_7^3 & . \\ . & . & \zeta_7^6 \\ \zeta_7^5 & . & . \end{pmatrix}$
	24	$(3,4,4)$	S_4	$\begin{pmatrix} . & . & 1 \\ 1 & . & . \\ . & 1 & . \end{pmatrix}, \begin{pmatrix} . & -1 & . \\ 1 & . & . \\ . & . & -1 \end{pmatrix}$
		$(2,6,6)$	$A_4 \times 2$	$\begin{pmatrix} 1 & . & . \\ . & -1 & . \\ . & . & -1 \end{pmatrix}, \begin{pmatrix} . & . & -1 \\ -1 & . & . \\ . & -1 & . \end{pmatrix}$
		$(2,2,2,3)$	S_4	$\begin{pmatrix} 1 & . & . \\ . & -1 & . \\ . & . & -1 \end{pmatrix}, \begin{pmatrix} . & . & -1 \\ -1 & . & . \\ . & -1 & . \end{pmatrix}, \begin{pmatrix} . & . & -1 \\ . & -1 & . \\ -1 & . & . \end{pmatrix}$

	48	$(3,3,4)$	$4^2:3$	$\begin{pmatrix} . & . & 1 \\ 1 & . & . \\ . & 1 & . \end{pmatrix},\ \begin{pmatrix} . & \zeta_4^3 & . \\ . & . & \zeta_4^3 \\ -1 & . & . \end{pmatrix}$
		$(2,4,6)$	$2\times S_4$	$\begin{pmatrix} . & -1 & . \\ -1 & . & . \\ . & . & -1 \end{pmatrix},\ \begin{pmatrix} . & . & 1 \\ . & -1 & . \\ -1 & . & . \end{pmatrix}$
	96	$(2,3,8)$	$4^2:S_3$	$\begin{pmatrix} -1 & . & . \\ . & . & -1 \\ . & -1 & . \end{pmatrix},\ \begin{pmatrix} . & . & 1 \\ \zeta_4^3 & . & . \\ . & \zeta_4 & . \end{pmatrix}$
	168	$(2,3,7)$	$L_3(2)$	$\begin{pmatrix} -1 & . & . \\ . & -1 & . \\ . & . & 1 \end{pmatrix},\ \tfrac{1}{2}\begin{pmatrix} 1 & -1 & -b_7^* \\ b_7 & b_7 & 0 \\ 1 & -1 & b_7^* \end{pmatrix}$
4	20	$(4,4,5)$	$5:4$	$\begin{pmatrix} . & . & . & 1 \\ . & . & 1 & . \\ 1 & . & . & . \\ . & 1 & . & . \end{pmatrix},\ \begin{pmatrix} . & . & \zeta_5 & . \\ . & . & . & \zeta_5^4 \\ . & \zeta_5^2 & . & . \\ \zeta_5^3 & . & . & . \end{pmatrix}$
	36	$(3,4,4)$	$3^2:4$	$\begin{pmatrix} \zeta_3 & . & . & . \\ . & \zeta_3^2 & . & . \\ . & . & \zeta_3 & . \\ . & . & . & \zeta_3^2 \end{pmatrix},\ \begin{pmatrix} . & . & . & 1 \\ . & . & 1 & . \\ . & 1 & . & . \\ 1 & . & . & . \end{pmatrix}$
		$(2,6,6)$	$S_3\times S_3$	$\begin{pmatrix} . & . & . & 1 \\ . & . & 1 & . \\ . & 1 & . & . \\ 1 & . & . & . \end{pmatrix},\ \begin{pmatrix} . & \zeta_3^2 & . & . \\ 1 & . & . & . \\ . & . & . & 1 \\ . & . & \zeta_3 & . \end{pmatrix}$
	60	$(2,5,5)$	A_5	$\begin{pmatrix} 1 & 1 & . & . \\ . & -1 & . & . \\ . & 1 & 1 & 1 \\ . & . & . & -1 \end{pmatrix},\ \begin{pmatrix} . & . & . & 1 \\ . & . & . & 1 \\ -1 & -1 & -1 & -1 \\ . & 1 & . & . \end{pmatrix}$
	72	$(2,4,6)$	$3^2:D_8$	$\begin{pmatrix} -1 & . & . & . \\ . & -1 & . & . \\ . & . & . & -1 \\ . & . & -1 & . \end{pmatrix},\ \begin{pmatrix} . & . & . & \zeta_6 \\ . & . & \zeta_6^5 & . \\ -1 & . & . & . \\ . & -1 & . & . \end{pmatrix}$
	120	$(2,4,5)$	S_5	$\begin{pmatrix} 1 & 1 & . & . \\ . & -1 & . & . \\ . & 1 & 1 & . \\ . & . & . & 1 \end{pmatrix},\ \begin{pmatrix} . & -1 & -1 & -1 \\ . & 1 & 1 & . \\ -1 & -1 & -1 & . \\ 1 & 1 & . & . \end{pmatrix}$
5	60	$(3,3,5)$	A_5	$\begin{pmatrix} . & . & 1 & . \\ . & \zeta_3^2 & . & . \\ 1 & . & . & . \\ . & . & . & \zeta_3 \end{pmatrix},\ \begin{pmatrix} \zeta_3^2 & . & . & . \\ . & . & \zeta_3^2 & . \\ . & . & . & \zeta_3 \\ . & \zeta_3^2 & . & . \end{pmatrix}$
	80	$(2,5,5)$	$2^4:5$	$\begin{pmatrix} 1 & . & . & . \\ . & -1 & . & . \\ . & . & -1 & . \\ . & . & . & -1 \end{pmatrix},\ \begin{pmatrix} . & . & . & 1 \\ 1 & . & . & . \\ . & 1 & . & . \\ . & . & 1 & . \end{pmatrix}$
	120	$(2,3,10)$	$A_5\times 2$	$\begin{pmatrix} -1 & . & . & . \\ . & . & . & \zeta_6 \\ . & . & -1 & . \\ . & \zeta_6^5 & . & . \end{pmatrix},\ \begin{pmatrix} . & 1 & . & . \\ . & . & \zeta_3 & . \\ \zeta_3^2 & . & . & . \\ . & . & . & \zeta_3^2 \end{pmatrix}$
	160	$(2,4,5)$	$2^4:D_{10}$	$\begin{pmatrix} . & . & . & -1 \\ . & . & -1 & . \\ . & -1 & . & . \\ -1 & . & . & . \end{pmatrix},\ \begin{pmatrix} -1 & . & . & . \\ . & . & . & -1 \\ . & . & -1 & . \\ . & 1 & . & . \end{pmatrix}$

6	72	$(2,4,9)$	$(2^2 \times 3).S_3$	$\begin{pmatrix} . & . & . & . & 1 & . \\ . & . & . & 1 & . & . \\ . & . & . & . & . & \zeta_3 \\ . & 1 & . & . & . & . \\ 1 & . & . & . & . & . \\ . & . & \zeta_3^2 & . & . & . \end{pmatrix},\ \begin{pmatrix} . & . & . & . & . & 1 \\ . & . & . & . & -1 & . \\ . & . & . & -1 & . & . \\ . & . & 1 & . & . & . \\ . & -1 & . & . & . & . \\ -1 & . & . & . & . & . \end{pmatrix}$
	120	$(2,4,6)$	S_5	$\begin{pmatrix} . & . & . & . & -1 & . \\ . & -1 & . & . & . & . \\ . & . & . & . & . & -1 \\ . & . & . & -1 & . & . \\ -1 & . & . & . & . & . \\ . & . & -1 & . & . & . \end{pmatrix},\ \begin{pmatrix} \zeta_4 & . & . & . & . & . \\ . & . & . & . & -1 & . \\ . & . & . & -1 & . & . \\ . & 1 & . & . & . & . \\ . & . & 1 & . & . & . \\ . & . & . & . & . & -\zeta_4 \end{pmatrix}$
7	56	$(2,7,7)$	$2^3:7$	$\begin{pmatrix} 1 & . & . & . & . & . & . \\ . & -1 & . & . & . & . & . \\ . & . & -1 & . & . & . & . \\ . & . & . & -1 & . & . & . \\ . & . & . & . & 1 & . & . \\ . & . & . & . & . & -1 & . \\ . & . & . & . & . & . & 1 \end{pmatrix},$
	504	$(2,3,7)$	$L_2(8)$	$\begin{pmatrix} . & . & . & . & . & . & 1 \\ 1 & . & . & . & . & . & . \\ . & 1 & . & . & . & . & . \\ . & . & 1 & . & . & . & . \\ . & . & . & 1 & . & . & . \\ . & . & . & . & 1 & . & . \\ . & . & . & . & . & 1 & . \end{pmatrix}\ \begin{pmatrix} 1 & . & . & . & . & . & . \\ . & 1 & . & . & . & . & . \\ . & . & -1 & . & . & . & . \\ . & . & . & 1 & . & . & . \\ . & . & . & . & -1 & . & . \\ . & . & . & . & . & -1 & . \\ . & . & . & . & . & . & -1 \end{pmatrix},$ $\tfrac{1}{2}\begin{pmatrix} . & . & -1 & 1 & 1 & . & 1 \\ . & -1 & 1 & 1 & . & 1 & . \\ 1 & -1 & -1 & . & -1 & . & . \\ -1 & -1 & . & -1 & . & . & 1 \\ 1 & . & 1 & . & . & -1 & 1 \\ . & -1 & . & . & 1 & -1 & -1 \\ 1 & . & . & -1 & 1 & 1 & . \end{pmatrix}$

8	168	$(3,3,4)$	$L_3(2)$	$\begin{pmatrix} \cdot & \zeta_3^2 & \cdot & \cdot & \cdot & \cdot & \cdot & \cdot \\ \cdot & \cdot & \cdot & \cdot & \cdot & \zeta_3^2 & \cdot & \cdot \\ \cdot & \cdot & \cdot & \cdot & \cdot & \cdot & \zeta_3 & \cdot \\ \cdot & \cdot & \zeta_3 & \cdot & \cdot & \cdot & \cdot & \cdot \\ \zeta_3^2 & \cdot & \cdot & \cdot & \cdot & \cdot & \cdot & \cdot \\ \cdot & \cdot & \cdot & \cdot & \zeta_3 & \cdot & \cdot & \cdot \\ \cdot & \cdot & \cdot & \cdot & \cdot & \cdot & \zeta_3^2 & \cdot \\ \cdot & \cdot & \cdot & \cdot & \cdot & \cdot & \cdot & \zeta_3 \end{pmatrix},$ $\begin{pmatrix} \cdot & \cdot & \cdot & \cdot & \cdot & \cdot & \cdot & 1 \\ \cdot & \cdot & \cdot & \cdot & \cdot & \zeta_3^2 & \cdot & \cdot \\ \cdot & \cdot & \cdot & \zeta_3^2 & \cdot & \cdot & \cdot & \cdot \\ \cdot & \cdot & \cdot & \cdot & \zeta_3 & \cdot & \cdot & \cdot \\ \cdot & \cdot & \cdot & \cdot & \cdot & \cdot & \zeta_3 & \cdot \\ \cdot & 1 & \cdot & \cdot & \cdot & \cdot & \cdot & \cdot \\ \zeta_3 & \cdot & \cdot & \cdot & \cdot & \cdot & \cdot & \cdot \\ \cdot & \cdot & \cdot & \cdot & \cdot & \cdot & \zeta_3^2 & \cdot \end{pmatrix}$
	336	$(2,3,8)$	$L_3(2).2$	$\begin{pmatrix} \cdot & \cdot & 1 & \cdot & \cdot & \cdot & \cdot & \cdot \\ \cdot & \cdot & \cdot & 1 & \cdot & \cdot & \cdot & \cdot \\ \cdot & \cdot & \cdot & \cdot & 1 & \cdot & \cdot & \cdot \\ 1 & \cdot & \cdot & \cdot & \cdot & \cdot & \cdot & \cdot \\ \cdot & 1 & \cdot & \cdot & \cdot & \cdot & \cdot & \cdot \\ \cdot & \cdot & 1 & \cdot & \cdot & \cdot & \cdot & \cdot \\ \cdot & \cdot & \cdot & \cdot & \cdot & -1 & \cdot & \cdot \\ \cdot & \cdot & \cdot & \cdot & \cdot & \cdot & -1 & \cdot \end{pmatrix},$ $\begin{pmatrix} \zeta_3 & \cdot & \cdot & \cdot & \cdot & \cdot & \zeta_6 & \cdot \\ \cdot & \cdot & \zeta_3^2 & \cdot & \cdot & \cdot & \cdot & \cdot \\ \cdot & \cdot & \cdot & \cdot & \zeta_6 & \cdot & \cdot & \cdot \\ \cdot & \cdot & \cdot & \cdot & \cdot & \zeta_6^5 & \cdot & \cdot \\ \cdot & \cdot & \cdot & \cdot & 1 & \cdot & \cdot & \cdot \\ \cdot & \cdot & \cdot & \cdot & \cdot & \cdot & \zeta_3^2 & \cdot \\ \cdot & \zeta_6 & \cdot & \cdot & \cdot & \cdot & \cdot & \cdot \end{pmatrix}$
10	360	$(2,4,5)$	A_6	$\begin{pmatrix} \cdot & \cdot & -1 & \cdot & \cdot & \cdot & \cdot & \cdot & \cdot & \cdot \\ \cdot & -1 & \cdot & \cdot & \cdot & \cdot & \cdot & \cdot & \cdot & \cdot \\ -1 & \cdot & \cdot & \cdot & \cdot & \cdot & \cdot & \cdot & \cdot & \cdot \\ \cdot & \cdot & \cdot & -1 & \cdot & \cdot & \cdot & \cdot & \cdot & \cdot \\ \cdot & \cdot & \cdot & \cdot & \cdot & \cdot & \cdot & \cdot & \zeta_4 & \cdot \\ \cdot & \cdot & \cdot & \cdot & \cdot & \cdot & \cdot & \cdot & \cdot & -\zeta_4 \\ \cdot & \cdot & \cdot & \cdot & \cdot & \cdot & \cdot & -\zeta_4 & \cdot & \cdot \\ \cdot & \cdot & \cdot & \cdot & \cdot & \cdot & \zeta_4 & \cdot & \cdot & \cdot \\ \cdot & \cdot & \cdot & \cdot & \cdot & -\zeta_4 & \cdot & \cdot & \cdot & \cdot \\ \cdot & \cdot & \cdot & \cdot & \zeta_4 & \cdot & \cdot & \cdot & \cdot & \cdot \end{pmatrix},$ $\begin{pmatrix} \cdot & \cdot & \cdot & \cdot & \cdot & \zeta_4 & \cdot & \cdot & \cdot & \cdot \\ \cdot & \cdot & \cdot & \cdot & \cdot & \cdot & -\zeta_4 & \cdot & \cdot & \cdot \\ 1 & \cdot & \cdot & \cdot & \cdot & \cdot & \cdot & \cdot & \cdot & -\zeta_4 \\ \cdot & \cdot & \cdot & \cdot & \cdot & \zeta_4 & \cdot & \cdot & \cdot & \cdot \\ \cdot & \cdot & \cdot & \cdot & \cdot & \cdot & \cdot & \cdot & -\zeta_4 & \cdot \\ \cdot & \cdot & -1 & \cdot & \cdot & \cdot & \cdot & \cdot & \cdot & \cdot \\ \cdot & \cdot & \cdot & \cdot & \cdot & \cdot & \cdot & -\zeta_4 & \cdot & \cdot \\ \cdot & \cdot & \cdot & -1 & \cdot & \cdot & \cdot & \cdot & \cdot & \cdot \\ \cdot & 1 & \cdot & \cdot & \cdot & \cdot & \cdot & \cdot & \cdot & \cdot \end{pmatrix}$

14	1 092	$(2,3,7)$	$L_2(13)$

$$
\begin{pmatrix}
\cdot & \cdot & \cdot & \cdot & \zeta_6 & \cdot & \cdot & \cdot & \cdot & \cdot & \cdot & \cdot & 1 & \cdot & \cdot \\
\cdot & \cdot & \cdot & \cdot & \cdot & \cdot & \cdot & \cdot & \zeta_3 & \cdot & \cdot & \cdot & \cdot & \cdot & \cdot \\
\cdot & \cdot & \cdot & \cdot & \cdot & \cdot & \zeta_3 & \cdot & \cdot & \cdot & \cdot & \cdot & \cdot & \cdot & \cdot \\
\zeta_6^5 & \cdot & \cdot & \cdot & \cdot & \cdot & \cdot & \cdot & \cdot & \cdot & \cdot & \cdot & \cdot & \cdot & \cdot \\
\cdot & \cdot & \cdot & \zeta_3^2 & \cdot & \cdot & \cdot & \cdot & \cdot & \cdot & \cdot & \cdot & \cdot & \cdot & \cdot \\
\cdot & \cdot & \cdot & \cdot & \cdot & \cdot & \cdot & \zeta_3 & \cdot & \cdot & \cdot & \cdot & \cdot & \cdot & \cdot \\
\cdot & \cdot & \cdot & \cdot & \cdot & \cdot & \zeta_3^2 & \cdot & \cdot & \cdot & \cdot & \cdot & \cdot & \cdot & \cdot \\
\cdot & \cdot & \zeta_3^2 & \cdot & \cdot & \cdot & \cdot & \cdot & \cdot & \cdot & \cdot & \cdot & \cdot & \cdot & \cdot \\
\cdot & \cdot & \cdot & \cdot & \cdot & \cdot & \cdot & \cdot & \cdot & \cdot & \zeta_3^2 & \cdot & \cdot & \cdot & \cdot \\
\cdot & \cdot & \cdot & \cdot & \cdot & \cdot & \cdot & \cdot & \cdot & \zeta_3 & \cdot & \cdot & \cdot & \cdot & \cdot \\
\cdot & 1 & \cdot & \cdot & \cdot & \cdot & \cdot & \cdot & \cdot & \cdot & \cdot & \cdot & \cdot & \cdot & \cdot \\
\cdot & \cdot & \cdot & \cdot & \cdot & \cdot & \cdot & \cdot & \cdot & \cdot & \cdot & \cdot & -1 & \cdot & \cdot \\
\cdot & \cdot & \cdot & \cdot & \cdot & \cdot & \cdot & \cdot & \cdot & \cdot & \cdot & \cdot & \cdot & -1 & \cdot
\end{pmatrix},
$$

$$
\begin{pmatrix}
\cdot & \cdot & \cdot & -1 & \cdot & \cdot & \cdot & \cdot & \cdot & \cdot & \cdot & \cdot & \cdot & \cdot & \cdot \\
\cdot & \cdot & \cdot & \cdot & \cdot & \cdot & \cdot & \cdot & \zeta_3^2 & \cdot & \cdot & \cdot & \cdot & \cdot & \cdot \\
\cdot & \cdot & \cdot & \cdot & \cdot & \cdot & \cdot & \cdot & \cdot & \cdot & \cdot & \cdot & \zeta_3 & \cdot & \cdot \\
\cdot & \cdot & \cdot & \cdot & \cdot & \cdot & \cdot & \zeta_6^5 & \cdot & \cdot & \cdot & \cdot & \cdot & \cdot & \cdot \\
\cdot & \cdot & \cdot & \cdot & \cdot & \cdot & \cdot & \cdot & \cdot & \zeta_3^2 & \cdot & \cdot & \cdot & \cdot & \cdot \\
\cdot & \cdot & \cdot & \cdot & \cdot & \cdot & \zeta_3^2 & \cdot & \cdot & \cdot & \cdot & \cdot & \cdot & \cdot & \cdot \\
\cdot & \cdot & \cdot & \zeta_6^5 & \cdot & \cdot & \cdot & \cdot & \cdot & \cdot & \cdot & \cdot & \cdot & \cdot & \cdot \\
\zeta_3^2 & \cdot & \cdot & \cdot & \cdot & \cdot & \cdot & \cdot & \cdot & \cdot & \cdot & \cdot & \cdot & \cdot & \cdot \\
\cdot & \cdot & \cdot & \cdot & \cdot & \cdot & \cdot & \cdot & \cdot & \cdot & \zeta_6 & \cdot & \cdot & \cdot & \cdot \\
\cdot & \cdot & \cdot & \cdot & \cdot & \cdot & \cdot & \cdot & \cdot & \cdot & \cdot & \cdot & \cdot & \zeta_3 & \cdot \\
\cdot & \zeta_6 & \cdot & \cdot & \cdot & \cdot & \cdot & \cdot & \cdot & \cdot & \cdot & \cdot & \cdot & \cdot & \cdot \\
\cdot & \cdot & \cdot & \cdot & \cdot & \cdot & \cdot & \cdot & \cdot & \cdot & \cdot & \zeta_3 & \cdot & \cdot & \cdot \\
\cdot & \cdot & \cdot & \cdot & \cdot & \cdot & \zeta_6^5 & \cdot & \cdot & \cdot & \cdot & \cdot & \cdot & \cdot & \cdot \\
\cdot & \cdot & \cdot & 1 & \cdot & \cdot & \cdot & \cdot & \cdot & \cdot & \cdot & \cdot & \cdot & \cdot & \cdot
\end{pmatrix}
$$

APPENDIX C

Maillet's Determinant

In this appendix we study how far the value of the expression

$$\sum_{\substack{1 \leq u \leq m \\ \gcd(u,m)=1}} f_u \frac{\zeta_m^u}{1 - \zeta_m^u},$$

for a positive integer m, determines the rational coefficients f_u. This question arises in Chapter 7.

A Number Theoretic Result

DEFINITION C.1. Let m be a positive integer, and $I(m)$ the set of prime residues modulo m. A subset I of $I(m)$ is called *complementary* if for each $u \in I(m)$, exactly one of u, $m - u$ is contained in I. We say that m has the property (LI) if there is a complementary subset I of $I(m)$ such that the set

$$\{1\} \cup \left\{ \frac{\zeta_m^u}{1 - \zeta_m^u} \,\middle|\, u \in I \right\}$$

is linearly independent over $\mathbb{Q}$.

Complementary subsets of $I(m)$ clearly exist if $m > 2$. In particular, $m = 2$ does *not* have the property (LI).

Because of the equality

$$(*) \qquad \frac{\zeta_m^u}{1 - \zeta_m^u} + \frac{\zeta_m^{-u}}{1 - \zeta_m^{-u}} = -1$$

shown in Lemma 12.2, we have

$$\sum_{u \in I(m)} f_u \frac{\zeta_m^u}{1 - \zeta_m^u} = \sum_{u \in I} (f_u - f_{m-u}) \frac{\zeta_m^u}{1 - \zeta_m^u} - \sum_{u \in I} f_{m-u}$$

for any complementary subset I of $I(m)$. So we can expect at most that the values $f_u - f_{m-u}$, for $u \in I$, and $\sum_{u \in I} f_{m-u}$ can be recovered. These values can in fact be recovered if and only if m has the property (LI). Furthermore, we see that all complementary subsets of $I(m)$ are equally good.

The main result of this section is

184

THEOREM C.2. *Each prime power larger than 2 has the property* (LI).

To prove this, we proceed in the following steps. First we construct a basis B of $\mathbb{Q}(\zeta_{p^k})$ for prime powers p^k. Then we show that the property (LI) is equivalent to the linear independence of the imaginary parts of the numbers $\zeta_m^u/(1 - \zeta_m^u)$, $u \in I$. Third, we compute the B-coefficients of these values, and finally we show that the coefficient matrix has full rank.

Throughout the chapter, for integers u with $\gcd(u, m) = 1$, let σ_u denote the field automorphism of $\mathbb{Q}(\zeta_m)$ that is defined by $\zeta_m^{\sigma_u} = \zeta_m^u$.

LEMMA C.3. *Let p be a prime, $k \geq 1$, and $b = \sum_{i=1}^k \zeta_{p^i}$. Then*

$$B = (b^{\sigma_u}; u \in I(p^k))$$

is a $\mathbb{Q}$-basis of $\mathbb{Q}(\zeta_{p^k})$.

Proof. We show that every p^k-th root of unity can be written as a linear combination of elements in B. For fixed j with $1 \leq j \leq k$, let G denote the Galois group $\mathrm{Aut}(\mathbb{Q}(\zeta_{p^k}), \mathbb{Q}(\zeta_{p^j}))$, and observe that the relative trace map

$$\mathrm{Tr}_{k,j} \colon \begin{array}{ccc} \mathbb{Q}(\zeta_{p^k}) & \to & \mathbb{Q}(\zeta_{p^j}) \\ \zeta_{p^k} & \mapsto & \sum_{\sigma \in G} \zeta_{p^k}^\sigma \end{array}$$

maps ζ_{p^i} to $p^{k-j}\zeta_{p^i}$ if $i \leq j$ and to 0 if $j < i \leq k$.

Thus $\mathrm{Tr}_{k,j}(b^{\sigma_u}) = p^{k-j}\sum_{i=1}^j \zeta_{p^i}^u$ for $j \in \{1, 2, \ldots, k\}$. Since B is closed under Galois conjugacy, the relative traces of b^{σ_u} are linear combinations of elements in B, and we get scalar multiples of all p-th, p^2-th, $\ldots$, p^k-th roots of unity as linear combinations of B. $\square$

LEMMA C.4. *Set $\eta_m = -m\,(1/2 + \zeta_m/(1 - \zeta_m))$, and let I be a complementary subset of $I(m)$. Then m has the property* (LI) *if and only if $(\eta_m^{\sigma_u}; u \in I)$ is linearly independent over $\mathbb{Q}$.*

Proof. The "only if" statement is clear. For the other direction, we only need to show that 1 is linearly independent of $(\eta_m^{\sigma_u}; u \in I)$ over $\mathbb{Q}$. But this follows from the fact that the $\eta_m^{\sigma_u}$ are purely imaginary, see equation $(*)$ above. $\square$

LEMMA C.5.

$$\frac{\zeta_m}{1 - \zeta_m} = \frac{-1}{m}\sum_{i=1}^m i\zeta_m^i = \frac{1}{m}\sum_{i=1}^{m-1} i\zeta_m^{-i}.$$

Proof. The first equality follows from

$$(1 - \zeta_m)\sum_{i=1}^m i\zeta_m^i = \sum_{i=1}^m i\zeta_m^i - \sum_{i=1}^m i\zeta_m^{i+1} = \sum_{i=1}^m \zeta_m^i - m\zeta_m^{m+1} = -m\zeta_m,$$

the second is a consequence of $1 = \zeta_m^m = \sum_{i=1}^{m-1} (-\zeta_m^i)$. □

LEMMA C.6. *For $u \in \mathbb{Z}$ with $(u, m) = 1$, define $\bar{u} \in I(m)$ and $R(u) \in I(m)$ by $u\bar{u} \equiv 1 \pmod{m}$ and $R(u) \equiv u \pmod{m}$, respectively. Let p, k, b be as above, let I be a complementary subset of $I(p^k)$, and let $\iota = \sigma_{-1}$ denote complex conjugation. Then we have*

(a) $\eta_{p^k} = \sum_{i \in I(p^k)} \left(i - \frac{p^k}{2} \right) b^{\sigma_i}$ *and*

(b) $\eta_{p^k}^{\sigma_u} = \sum_{j \in I} \left(R(j\bar{u}) - \frac{p^k}{2} \right) (b - b^\iota)^{\sigma_i}$.

Proof. By Lemma C.5, we have

$$\eta_m = -\frac{m}{2} + \sum_{i=1}^{m} i\zeta_m^i = -\frac{m}{2} + \sum_{i=1}^{m} \left(i - \frac{m}{2} \right) \zeta_m^i = \sum_{i=1}^{m-1} \left(i - \frac{m}{2} \right) \zeta_m^i.$$

For prime powers $m = p^k$, this is equal to $\sum_{i \in I(p^k)} \left(i - \frac{p^k}{2} \right) \sum_{j=1}^{k} \zeta_{p^j}^i$ if and only if

$$\sum_{\substack{1 \leq i \leq p^k - 1 \\ p | i}} \left(i - \frac{p^k}{2} \right) \zeta_{p^k}^i = \sum_{\substack{1 \leq i \leq p^k - 1 \\ p \nmid i}} \left(i - \frac{p^k}{2} \right) \sum_{j=1}^{k-1} \zeta_{p^j}^i.$$

For the summands on the left hand side, the index i is of the form $i = p^{k-j}i'$ with $k - j \geq 1$ and $p \nmid i'$, so the left hand side is equal to

$$\sum_{j=1}^{k-1} \sum_{\substack{1 \leq i' \leq p^j - 1 \\ p \nmid i'}} \left(p^{k-j}i' - \frac{p^k}{2} \right) \zeta_{p^k}^{p^{k-j}i'}.$$

The multiplicity of $\zeta_{p^k}^{p^{k-j}i'} = \zeta_{p^j}^{i'}$ in this sum is $p^{k-j}i' - \frac{p^k}{2}$, its multiplicity in the sum on the right hand side is

$$\sum_{\substack{1 \leq i \leq p^k - 1 \\ i \equiv i' \pmod{p^j}}} \left(i - \frac{p^k}{2} \right) = \sum_{l=0}^{p^{k-j}-1} \left((i' + lp^j) - \frac{p^k}{2} \right)$$

$$= p^{k-j}i' + p^j \frac{p^{k-j}(p^{k-j} - 1)}{2} - p^{k-j}\frac{p^k}{2} = p^{k-j}i' - \frac{p^k}{2}.$$

This proves part (a). To deduce part (b) from (a), consider the equality

$$\eta_{p^k}^{\sigma_u} = \sum_{i \in I(p^k)} \left(i - \frac{p^k}{2} \right) b^{\sigma_i \sigma_u} = \sum_{j \in I(p^k)} \left(R(j\bar{u}) - \frac{p^k}{2} \right) b^{\sigma_j},$$

which holds because $j = R(iu)$ runs over $I(m)$ as i does, and $i = R(i) = R(iu\bar{u}) = R(j\bar{u})$. Now $R((m-j)\bar{u}) - \frac{m}{2} = R(m - j\bar{u}) - \frac{m}{2} = m - R(j\bar{u}) - \frac{m}{2} = -\left(R(j\bar{u}) - \frac{m}{2} \right)$ yields the desired expression. □

By Lemma C.6 (b), the statement of Theorem C.2 for the case that $m = p^k$ is a prime power is equivalent to

$$\det \left(R(j\bar{u}) - \frac{p^k}{2} \right)_{u,j \in I} \neq 0.$$

Thus Theorem C.2 follows from

THEOREM C.7 (Carlitz and Olson 1955, Kühnová 1979, Tateyama 1982). *If* p, k, I *are as in Lemma C.6 then*

$$\det \left(R(j\bar{u}) - \frac{p^k}{2} \right)_{u,j \in I} = -(-p^k)^{\varphi(p^k)/2} \frac{h^-_{p^k}}{w_{p^k}},$$

where h^-_m *and* w_m *denote the relative class number of* $\mathbb{Q}(\zeta_m)$ *and the number of roots of unity in* $\mathbb{Q}(\zeta_m)$, *respectively.*

For the proof of this theorem, we need a nontrivial result from number theory and a technical lemma.

THEOREM C.8. *Let* h^-_m *and* w_m *be as above, and let* Q_m *denote the unit index of* $\mathbb{Q}(\zeta_m)$. *Set*

$$B_{1,\chi} = \frac{1}{m} \sum_{a=1}^{m} \chi(a) a$$

for each nontrivial irreducible character χ *of* $(\mathbb{Z}/m\mathbb{Z})^\times$, *where* $\chi(a) = 0$ *for* $(a, m) \neq 1$. *Then*

$$h^-_m = Q_m w_m \prod_\chi \left(-\tfrac{1}{2} B_{1,\chi} \right),$$

where the product is taken over all odd *irreducible characters* χ *of* $(\mathbb{Z}/m\mathbb{Z})^\times$, *i.e., those* χ *with* $\chi(-1) = -1$.

If m *is a prime power then* $Q_m = 1$, *otherwise* $Q_m = 2$.

Proof. The statement about h^-_m is [**Was96**, Theorem 4.17]. The $B_{1,\chi}$ are defined on p. 31 of [**Was96**], their expression given above is stated on page 32 of [**Was96**]. The statement about Q_m is [**Was96**, Corollary 4.13]. □

LEMMA C.9 (see [**Was96**, Lemma 5.26 (a)]). *Let* G *be a finite abelian group and let* f *be a function on* G *with values in some field* K *of characteristic 0. Then*

$$\det \left(f(\sigma \tau^{-1}) \right)_{\sigma, \tau \in G} = \prod_{\chi \in \mathrm{Irr}(G)} \sum_{\sigma \in G} f(\sigma) \chi(\sigma).$$

Proof. Let $\rho\colon G \to GL_{|G|}(K)$ be the right regular representation of the finite group G, i.e., for $\sigma \in G$ the rows and columns of the matrix $\rho(\sigma)$ are indexed by the elements of G, the entry $\rho(\sigma)_{\alpha,\beta}$ in row α and column β being 1 if $\alpha\sigma = \beta$, and 0 otherwise. Then

$$\left(f(\alpha^{-1}\beta)\right)_{\alpha,\beta\in G} = \sum_{\sigma\in G} f(\sigma)\rho(\sigma),$$

because $\sum_{\sigma\in G} f(\sigma)\rho(\sigma)_{\alpha,\beta} = f(\alpha^{-1}\beta)$.

If G is abelian then the $\rho(\sigma)$, $\sigma \in G$, can be conjugated simultaneously to diagonal matrices with diagonal entries $\chi(\sigma)$, $\chi \in \mathrm{Irr}(G)$. Thus

$$\begin{aligned}
\det\left(f(\alpha^{-1}\beta)\right)_{\alpha,\beta\in G} &= \det\left(\sum_{\sigma\in G} f(\sigma)\rho(\sigma)\right)\\
&= \det\left(\sum_{\sigma\in G} f(\sigma)\mathrm{diag}(\chi(\sigma) \mid \chi \in \mathrm{Irr}(G))\right)\\
&= \prod_{\chi\in\mathrm{Irr}(G)} \sum_{\sigma\in G} f(\sigma)\chi(\sigma),
\end{aligned}$$

and the statement follows by reordering of rows and columns of the matrix. $\square$

Now we prove Theorem C.7.

Proof. (The proof follows [**CO55**].) We identify $I(p^k)$ with the group $(\mathbb{Z}/p^k\mathbb{Z})^\times$, and first show that

$$(*) \qquad \det\left(R(j\overline{u}) - \frac{p^k}{2}\right)_{u,j\in I} = \prod_{\substack{\chi\in\mathrm{Irr}(I(p^k))\\ \mathrm{odd}}} \frac{1}{2} \sum_{u\in I(p^k)} u \cdot \chi(u).$$

Now we choose an odd character $\chi_0 \in \mathrm{Irr}(I(p^k))$, and conjugate our matrix with the diagonal matrix $\mathrm{diag}(\chi_0(u) \mid u \in I)$, which yields

$$\mathrm{diag}(\chi_0(\overline{u}) \mid u \in I) \cdot \left(R(j\overline{u}) - \frac{p^k}{2}\right)_{u,j\in I} \cdot \mathrm{diag}(\chi_0(j) \mid j \in I)$$

$$= \left(\left(R(j\overline{u}) - \frac{p^k}{2}\right) \cdot \chi_0(j\overline{u})\right)_{u,j\in I}.$$

Setting $f(a) = \left(a - \frac{p^k}{2}\right)\chi_0(a)$, we have $f(p^k-a) = f(a)$, so f can be regarded as a function on the factor group $G = I(p^k)/\{\pm 1\}$. Applying Lemma C.9, we get

$$
\det\left(R(j\bar{u}) - \frac{p^k}{2}\right)_{u,j\in I} = \det\left(\left(R(j\bar{u}) - \frac{p^k}{2}\right)\cdot\chi_0(j\bar{u})\right)_{u,j\in I}
$$

$$
= \prod_{\chi\in\mathrm{Irr}(G)}\sum_{u\in I}\left(u - \frac{p^k}{2}\right)\cdot\chi_0(u)\cdot\chi(u),
$$

and calculate

$$
\sum_{u\in I}\left(2u - p^k\right)\cdot\chi_0(u)\cdot\chi(u)
$$

$$
= \sum_{u\in I}u\cdot\chi_0(u)\cdot\chi(u) + \sum_{u\in I}(p^k - u)\cdot\chi_0(p^k - u)\cdot\chi(p^k - u)
$$

$$
= \sum_{u\in I(p^k)}u\cdot\chi_0(u)\cdot\chi(u).
$$

Since $\chi_0\cdot\chi$ is an odd character of $I(p^k)$, formula $(*)$ follows.

Now $\sum_{u\in I(p^k)}u\cdot\chi(u) = \sum_{i=1}^{p^k-1}i\cdot\chi(i)$ if χ is a nontrivial character, because $\chi(i) = 0$ if i is divisible by p. Thus

$$
\det\left(R(j\bar{u}) - \frac{p^k}{2}\right)_{u,j\in I} = \prod_{\substack{\chi\in\mathrm{Irr}(I(p^k))\\ \mathrm{odd}}}\frac{1}{2}\sum_{i=1}^{p^k-1}i\cdot\chi(i) = \prod_{\substack{\chi\in\mathrm{Irr}(I(p^k))\\ \mathrm{odd}}}\frac{1}{2}p^k B_{1,\chi}
$$

$$
= (-p^k)^{\varphi(p^k)/2}\frac{h^-_{p^k}}{Q_{p^k}w_{p^k}} = (-p^k)^{\varphi(p^k)/2}\frac{h^-_{p^k}}{w_{p^k}},
$$

as desired.   □

Some Remarks on the Problem

In order to prove the linear independence of $(\eta_m^{\sigma_u}; u\in I)$ for a prime power m, we need not compute the determinant of the matrix $\left(R(j\bar{u}) - \frac{m}{2}\right)_{u,j\in I}$. It would be sufficient to prove that the values $B_{1,\chi}$ are nonzero for odd χ. On page 31 of [**Was96**] it is stated that no elementary proof of this fact is known. Its proof in [**Was96**] uses the expression for h^-_m given in Theorem C.8.

The term "Maillet's Determinant", although mentioned in the title of the appendix, has not been explained yet. In order to do this now, we write

$$
D_m = \left(R(a\bar{b})\right)_{a,b\in I}
$$

for any complementary subset I of $I(m)$. (Note that if the determinant is nonzero for *one* complementary subset then its value does not depend on the choice of I, and otherwise $D_m = 0$ holds for all complementary subsets of

$I(m)$.) Maillet conjectured that for a prime p, $\det D_p \neq 0$. Carlitz and Olson proved this in [**CO55**] by calculating

$$\det D_p = \pm p^{(p-3)/2} h_p^-.$$

This theorem is generalized in [**Küh79**] to

THEOREM C.10 (Kühnová). *If m is a prime power then*

$$\det D_m = (-m)^{\varphi(m)/2-1} h_m^-.$$

Proof. (see [**CO55**]) We choose $I = \{u \in I(m) \mid 1 \leq u < m/2\}$, and consider the matrix $\left(R(a\bar{b}) + x\right)_{a,b\in I}$. For odd m, the last column is that of $b = \frac{m-1}{2}$, it consists of the values $m - 2a + x$ since $\frac{m-1}{2}(m - 2a) \equiv a \pmod{m}$. Adding the first column twice to the last yields the values $m + 3x$, so the indeterminate can be eliminated by subtraction of the $\frac{x}{m+3x}$-fold multiple of the last column. This yields

$$\det \left(R(a\bar{b}) + x\right)_{a,b\in I} = \frac{m + 3x}{m} \det \left(R(a\bar{b})\right)_{a,b\in I}.$$

Setting $x = -\frac{m}{2}$, we get $\det \left(R(a\bar{b}) - \frac{m}{2}\right)_{a,b\in I} = -\frac{1}{2}\det D_m$.

For $m \geq 4$ even, the last column belongs to $b = \frac{m}{2} - 1$, it consists of the values $\frac{m}{2} - a + x$ since $\left(\frac{m}{2} - 1\right)\left(\frac{m}{2} - a\right) = m\frac{m/2-(a+1)}{2} + a \equiv a \pmod{m}$. Using manipulations as above, we get

$$\det \left(R(a\bar{b}) + x\right)_{a,b\in I} = \frac{m + 4x}{m} \det \left(R(a\bar{b})\right)_{a,b\in I}$$

and thus $\det \left(R(a\bar{b}) - \frac{m}{2}\right)_{a,b\in I} = -\det D_m$.

Now the statement of the theorem follows from Theorem C.7 and the fact that $w_m = 2m$ if m is odd, and $w_m = m$ if m is even.   □

With the help of additional techniques, in [**Tat82**] this result is generalized to

THEOREM C.11 (Tateyama). *For any positive integer m, $m \not\equiv 2 \pmod 4$, we have*

$$\det D_m = \begin{cases} (-m)^{\varphi(m)/2-1} 2^{g(m)-1} h_m^-, & g(m) = g^+(m), \\ 0, & g(m) \neq g^+(m), \end{cases}$$

where $g(m)$ and $g^+(m)$ are the numbers of prime factors of the ideal spanned by m in the rings of integers in $\mathbb{Q}(\zeta_m)$ and $\mathbb{Q}(\zeta_m) \cap \mathbb{R}$, respectively.

So besides the relative class number h_m^- as a factor in the determinant of D_m, nontrivial number theory is involved in the answer to the completely elementary question whether the determinant of D_m, which is just one quarter of the multiplication table of $(\mathbb{Z}/m\mathbb{Z})^\times$, is nonzero.

How does Tateyama's Theorem affect the question whether a number m has the property (LI)? For example, we have $D_{20} = 0$, but 20 has the property (LI). In fact, all integers from 3 to 2000 have the property (LI), as has been shown using the computer algebra system GAP [S$^+$94], by computing the rank of the integral matrix $(2 \cdot (ij \pmod{m})) - m)_{i \in I, 1 \leq j \leq (m-1)/2}$, for a complementary subset I of $I(m)$. So we state

CONJECTURE C.12. *Each positive integer larger than 2 has the property* (LI).

Tateyama's Theorem shows first that it is hopeless to look for a nice normal basis as in Lemma C.3 if m is not a prime power. As a second consequence, the proof of the theorem shows that it is hopeless to look for a basis such that the determinant of the coefficient matrix is a product of terms of the form $\sum_{a \in I(m)} a\chi(a)$, for nontrivial characters χ; the point is that in general this sum may be zero, in particular it is not equal to $B_{1,\chi} = \sum_{i=1}^{m-1} i\chi(i)$.

Following the approach of Skula and Kučera in [**Sku81, Kuč92a, Kuč92b**], we can give a basis independent interpretation of $|\det(R(a\bar{b}) - \frac{m}{2})_{a,b \in I}|$. For that, we view $\mathbb{Z}[\zeta_m]$, the ring of integers in $\mathbb{Q}(\zeta_m)$, as a module for the Galois group $G = \text{Aut}(\mathbb{Q}(\zeta_m), \mathbb{Q}) = \{\sigma_u \mid u \in I(m)\}$. Let $\mathbb{Z}[\zeta_m]^-$ denote the eigenspace for σ_{-1} to the eigenvalue -1. We have $2\eta_m^{\sigma_u} \in \mathbb{Z}[\zeta_m]^-$ for $u \in I(m)$, and the linear independence of $(2\eta_m^{\sigma_u}; u \in I)$ means that the $\mathbb{Z}$-span L_η of these values is a lattice of finite index in $\mathbb{Z}[\zeta_m]^-$.

If $m = p^k$ is a prime power and b is as in Lemma C.3 then the index of the $\mathbb{Z}$-span L_b of $((b - b^{\sigma-1})^{\sigma_u}; u \in I)$ in $\mathbb{Z}[\zeta_m]$ is a power of p. Lemma C.6 implies that L_η is contained in L_b, and by Theorem C.7, the index of L_η in L_b is

$$(2p^k)^{\varphi(p^k)/2} \frac{h_{p^k}^-}{2w_{p^k}}.$$

Skula and Kučera deal with the integral group ring $\mathbb{Z}[G]$ instead of its (regular) module $\mathbb{Q}(\zeta_m)$. Skula constructs a sublattice of index $h_{p^k}^-$ in $\mathbb{Z}[G]^-$, the eigenspace of σ_{-1} for the eigenvalue -1 in $\mathbb{Z}[G]$.

Kučera generalizes this to arbitrary positive integers m. He constructs two lattices A, S in $\mathbb{Z}[G]^- \oplus \mathbb{Z} \cdot \sum_{u \in I} \sigma_u$ such that the index of S in A is h_m^- times a power of 2. Moreover, for the calculation of the index of S in A, he computes the determinant of the *regular* matrix

$$\left(R(j\bar{u}) - \frac{m}{2}\right)_{j \in I'(m), u \in M_-}$$

where $I'(m)$ is the complementary subset $\{u \in I(m) \mid 1 \leq u < m/2\}$ and M_- is a suitable subset of $\{0, 1, \ldots, m-1\}$. If m is a prime power then $M_- = I'(m)$.

References

[Acc68] R. D. M. Accola, *On the number of automorphisms of a closed Riemann surface*, Trans. Amer. Math. Soc. **131** (1968), 398–408.

[Acc94] R. D. M. Accola, *Topics in the Theory of Riemann Surfaces*, Lecture Notes in Mathematics, vol. 1595, Springer-Verlag, Berlin Heidelberg, 1994, ix+105 pp., ISBN 3-540-58721-7.

[Ahl54] L. V. Ahlfors, *On quasiconformal mappings*, L. Analyse Math. **3** (1954), 1–58.

[BE99] H.-U. Besche and B. Eick, *The groups of order at most 1000 except 512 and 768*, J. Symbolic Comput. **27** (1999), no. 4, 405–413.

[BEGG90] E. Bujalance, J. J. Etayo, J. M. Gamboa, and G. Gromadzki, *Automorphism Groups of Compact Bordered Klein Surfaces*, Lecture Notes in Mathematics, vol. 1439, Springer-Verlag, Berlin Heidelberg New York, 1990, xiii+201 pp., ISBN 3-540-52941-1.

[Bon93] O. Bonten, *Über Kommutatoren in endlichen einfachen Gruppen*, Dissertation, Lehrstuhl D für Mathematik, Rheinisch-Westfälische Technische Hochschule Aachen, 1993.

[BP98] Thomas Breuer and Götz Pfeiffer, *Finding possible permutation characters*, J. Symbolic Comput. **26** (1998), 343–354.

[Bur79] R. Burkhardt, *Über die Zerlegungszahlen der Suzukigruppen $Sz(q)$*, J. Algebra **59** (1979), 421–433.

[CCN+85] J. H. Conway, R. S. Curtis, S. P. Norton, R. A. Parker, and R. A. Wilson, *An Atlas of Finite Groups*, Clarendon Press, Oxford, 1985, xxxiii+252 pp., ISBN 0-19-853199-0.

[CO55] L. Carlitz and F. R. Olson, *Maillet's determinant*, Proc. Amer. Math. Soc. **6** (1955), 265–269.

[Con85] M. D. E. Conder, *The symmetric genus of alternating and symmetric groups*, J. Combin. Theory Ser. B (1985), no. 39, 179–186.

[Con87] M. D. E. Conder, *The genus of compact Riemann surfaces with maximal automorphism group*, J. Algebra **108** (1987), 204–247.

[Con91] M. D. E. Conder, *The symmetric genus of the Mathieu groups*, Bull. London Math. Soc. (1991), no. 23, 445–453.

[CW34] C. Chevalley and A. Weil, *Über das Verhalten der Integrale 1. Gattung bei Automorphismen des Funktionenkörpers*, Abh. Math. Sem. Univ. Hamburg **10** (1934), 358–361.

[CWW92] M. D. E. Conder, R. A. Wilson, and A. J. Woldar, *The symmetric genus of sporadic groups*, Proc. Amer. Math. Soc. **116** (1992), no. 3, 653–663.

[FK92] H. M. Farkas and I. Kra, *Riemann Surfaces*, 2nd ed., Graduate Texts in Mathematics, vol. 71, Springer-Verlag, Berlin Heidelberg New York, 1992, xvi+363 pp., ISBN 3-540-97703-1.

[For77] O. Forster, *Riemannsche Flächen*, Heidelberger Taschenbücher, vol. 184, Springer-Verlag, Berlin Heidelberg, 1977, x+223 pp., ISBN 3-540-08034-1.

[Gre63] L. Greenberg, *Maximal Fuchsian groups*, Bull. Amer. Math. Soc. **69** (1963), 569–573.

[Gue82] I. Guerrero, *Holomorphic families of compact riemann surfaces with automorphisms*, Illinois J. Math. **26** (1982), no. 2, 212–225.

[Har66] W. J. Harvey, *Cyclic groups of automorphisms of a compact Riemann surface*, Quart. J. Math. Oxford Ser. (2) **17** (1966), 86–97.

[Har71] W. J. Harvey, *On branch loci in Teichmüller space*, Trans. Amer. Math. Soc. **153** (1971), 387–399.

[HP89] D. F. Holt and W. Plesken, *Perfect Groups*, Oxford University Press, 1989, xii+364 pp., ISBN 0-19-853559-7.

[Hup83] B. Huppert, *Endliche Gruppen I*, Grundlehren der mathematischen Wissenschaften, vol. 134, Springer-Verlag, Berlin Heidelberg, 1983, xvi+796 pp., ISBN 3-540-03825-6.

[Hur93] A. Hurwitz, *Über algebraische Gebilde mit eindeutigen Transformationen in sich*, Math. Ann. **41** (1893), 403–442.

[Isa76] I. M. Isaacs, *Character Theory of Finite Groups*, Pure and Applied Mathematics, vol. 69, Academic Press, New York, 1976, xii+303 pp., ISBN 0-12-374550-0.

[Isa77] I. M. Isaacs, *Commutators and the commutator subgroup*, Amer. Math. Monthly **84** (1977), 720–722.

[Itô51] N. Itô, *A theorem on the alternating group A_n ($n \geq 5$)*, Math. Japon. **2** (1951), 59–60.

[JK81] G. D. James and A. Kerber, *The Representation Theory of the Symmetric Group*, Encyclopedia of Mathematics and its Applications, vol. 16, Addison-Wesley, 1981, xxviii+510 pp., ISBN 0-201-13515-9.

[Jon95] G. A. Jones, *Enumeration of Homomorphisms and Surface-Coverings*, Quart. J. Math. Oxford Ser. (2) **46** (1995), 485–507.

[JS87] G. A. Jones and D. Singerman, *Complex functions: An algebraic and geometric viewpoint*, Cambridge University Press, 1987, xiv+342 pp., ISBN 0-521-31366-X.

[Kim93] H. Kimura, *Automorphism groups, isomorphic to D_8 or Q_8, of compact Riemann surfaces*, J. Pure Appl. Algebra **87** (1993), 23–36.

[KK90a] A. Kuribayashi and H. Kimura, *Automorphism groups of compact Riemann surfaces of genus five*, J. Algebra **134** (1990), 80–103.

[KK90b] I. Kuribayashi and A. Kuribayashi, *Automorphism groups of compact Riemann surfaces of genera three and four*, J. Pure Appl. Algebra **65** (1990), 277–292.

[KK91] A. Kuribayashi and H. Kimura, *Automorphism groups, isomorphic to $GL(3, \mathbf{F}_2)$, of compact Riemann surfaces*, Tôhoku Math. J. **43** (1991), 337–353.

[KPW87] P. B. Kleidman, R. A. Parker, and R. A. Wilson, *The maximal subgroups of the Fischer group Fi_{23}*, J. London Math. Soc. (2) **39** (1987), 89–101.

[Kuč92a] R. Kučera, *On bases of odd and even universal ordinary distributions*, J. Number Theory **40** (1992), 264–283.

[Kuč92b] R. Kučera, *On bases of the Stickelberger ideal and of the group of circular units of a cyclotomic field*, J. Number Theory **40** (1992), 264–283.

[Küh79] J. Kühnová, *Maillet's determinant $D_{p^{n+1}}$*, Arch. Math. (Brno) **15** (1979), no. 4, 209–212.

[Kur66] A. Kuribayashi, *On analytic families of compact Riemann surfaces with nontrivial automorphisms*, Nagoya Math. J. **28** (1966), 119–165.

[Kur83] I. Kuribayashi, *On automorphisms of prime order of a Riemann surface as matrices*, Manuscripta Math. **44** (1983), 103–108.

[Kur84] I. Kuribayashi, *On an algebraization of the Riemann–Hurwitz relation*, Kodai Math. J. **7** (1984), 222–237.

194 REFERENCES

[Kur86]   I. Kuribayashi, *Classification of automorphism groups of compact Riemann surfaces of genus two*, preprint, Tsukuba (1986), 25–39.

[Kur87]   I. Kuribayashi, *On automorphism groups of a curve as linear groups*, J. Math. Soc. Japan **39** (1987), no. 2, 51–77.

[Lan70]   S. Lang, *Algebraic Number Theory*, Addison-Wesley, 1970, xi+354 pp.

[Leh64]   J. Lehner, *Discontinuous Groups and Automorphic Functions*, Mathematical Surveys and Monographs, no. 8, American Mathematical Society, Providence, Rhode Island, 1964, xi+425 pp., ISBN 0-8218-1508-3.

[Lew63]   J. Lewittes, *Automorphisms of compact Riemann surfaces*, Amer. J. Math. **85** (1963), 734–752.

[Mac65]   C. Maclachlan, *Abelian groups of automorphisms of compact Riemann surfaces*, Proc. London Math. Soc. (3) **15** (1965), 699–712.

[Mac69]   C. Maclachlan, *A bound for the number of automorphisms of a compact Riemann surface*, J. London Math. Soc. (2) **44** (1969), 265–272.

[Mas71]   B. Maskit, *On Poincaré's theorem for fundamental polygons*, Adv. in Math. **7** (1971), 219–230.

[Mas91]   W. S. Massey, *A Basic Course in Algebraic Topology*, Graduate Texts in Mathematics, vol. 127, Springer-Verlag, Berlin Heidelberg New York, 1991, xvi+428 pp., ISBN 0-387-97430-X.

[Mat87]   B. H. Matzat, *Konstruktive Galoistheorie*, Lecture Notes in Mathematics, vol. 1284, Springer-Verlag, Berlin Heidelberg New York, 1987, x+286 pp., ISBN 3-540-18444-9.

[Mir95]   R. Miranda, *Algebraic Curves and Riemann Surfaces*, Graduate Studies in Mathematics, vol. 5, American Mathematical Society, Providence, Rhode Island, 1995, xxi+390 pp., ISBN 0-8218-0268-2.

[NO89]   M. F. Newman and E. A. O'Brien, *A CAYLEY library for the groups of order dividing 128*, Group Theory, Proceedings of the 1987 Singapore Conference, Walter de Gruyter, Berlin New York, 1989, pp. 437–442.

[O'Br91]   E. A. O'Brien, *The groups of order 256*, J. Algebra **143** (1991), 219–235.

[Rin94]   M. Ringe, *The C-MeatAxe, Version 2.04a*, Manual, 1994.

[S+94]   M. Schönert et al., *GAP – Groups, Algorithms and Programming*, Lehrstuhl D für Mathematik, RWTH Aachen, 3.4 ed., 1994.

[Sah69]   C.-H. Sah, *Groups related to compact Riemann surfaces*, Acta Math. **123** (1969), 13–42.

[Sch07]   I. Schur, *Untersuchungen über die Darstellung der endlichen Gruppen durch gebrochene lineare Substitutionen*, J. Math. **132** (1907), 85–137.

[Sco77]   L. L. Scott, *Matrices and cohomology*, Ann. of Math. **105** (1977), 473–492.

[Sku81]   L. Skula, *Another proof of Iwasawa's class number formula*, Acta Arith. **39** (1981), 1–6.

[Tat82]   K. Tateyama, *Maillet's determinant*, Sci. Papers College Gen. Edu. Univ. Tokyo **32** (1982), 97–100.

[Tuc83]   T. W. Tucker, *Finite groups acting on surfaces and the genus of a group*, J. Combin. Theory Ser. B **34** (1983), 82–98.

[Was96]   L. C. Washington, *Introduction to Cyclotomic Fields*, second ed., Graduate Texts in Mathematics, vol. 83, Springer-Verlag, Berlin Heidelberg New York, 1996, xiv+487 pp., ISBN 0-387-94762-0.

[Wim95]   A. Wiman, *Über die hyperelliptischen Curven und diejenigen vom Geschlechte p = 3, welche eindeutige Transformationen in sich zulassen*, Bihang Kongl. Svenska Vetenskaps–Akademiens Handlingar (1895), no. 21 (1), 1–23.

[Wol89a]   A. J. Woldar, *Genus actions of finite simple groups*, Illinois J. Math. **33** (1989), no. 3, 438–450.

[Wol89b] A. J. Woldar, *On Hurwitz generation and genus actions of sporadic groups*, Illinois J. Math. **33** (1989), no. 3, 416–437.

Index

$G\colon H$, XI
$[G\colon H]$, XI
$[\chi, \psi]$, 23
1_G, 23

A_n, *see* alternating group
abelian group, VII, 12, 24, 29–35, 44, 49,
   62, 64, 65, 73, 105, 106, 108, 111,
   119, 129, 154, 159–160, 176, 187
abelian invariants, 29–31, 49, 73, 176–178
Accola, R. D. M., 6, 90
admissible signature, 64–70
admissible system, 144
Ahlfors, L. V., 21
alternating group, XI, 62, 130, 133–137,
   163, 173
ambivalent group, 92
Atlas of Finite Groups, XII, 28, 40, 43,
   54, 56, 59, 61, 84, 89, 100
$\mathrm{Aut}(G)$, 2
$\mathrm{Aut}(X)$, 2
$\mathrm{Aut}_Y(X)$, 5
automorphism
   of a character table, 28
   of a group, 2
   of a Riemann surface, 2
   order of, 32
automorphism group
   abelian, 29–35, 159–160
   cyclic, 31
   full, 2, 3, 20–21

Besche, H.-U., VIII, 72, 73
branch point, 4
Bujalance, E., X
Burkhardt, R., 173

$\mathbb{C}$, 3
$\hat{\mathbb{C}}$, 3
$c.l.(G)$, 105
Carlitz, L., X, 187, 188, 190
centralizer, XI

character, 22
   faithful, 22, 41
   irreducible, 23, 81–84, 179–183
   linear, 24, 103, 151
   rational, 25
   rational irreducible, 25
   real, 25, 41, 92
   regular, 24
   trivial, 23
   virtual, 23
character table, 23
class function, 23
class multiplication coefficient, 47
class number, relative, 187
class structure, 46
complementary subset, 184
Conder, M. D. E., VII, IX, 47, 48, 54, 56,
   57, 62, 63
conformal automorphism, 2
conjugacy, XI
Conway, J. H., XII, 28, 40, 43, 54, 56, 59,
   61, 84, 89, 100
covering
   branched, 4
   regular, 5
   smooth, 4
   unbranched, 4
   universal, 4
covering transformation, 5
Curtis, R. S., *see* Conway, J. H.
(CY), 159
$CY(G)$, 35
$\widehat{CY}(G)$, 102
$CY(G, H)$, 35
$cy(G, h)$, 38

$\mathcal{D}$, 3
D_{2n}, *see* dihedral group
degree of a character, 22
differential form, 25–27
   invariant, 27

196